Praise for *Stem Cell Wars*

"An excellent account . . . the first of its kind."—*Cure Paralysis Now*

"An important new book."—*Crossleft*

"Comprehensive and concise, *Stem Cell Wars* provides an indispensable primer for anyone interested in what promises to be the most significant medical science breakthrough in our lifetimes. It should also serve as a timely antidote to the politically inspired misinformation surrounding this important issue."

—Ron Reagan

"Eve Herold is a latter-day Edward R. Murrow. . . . She's everywhere at once: behind closed doors on Capitol Hill, beside the scientists and the suffering patients they hope to save, even to South Korea where a fraud of historic proportions threatened to end the great promise of regenerative medicine. Her sympathies are unwaveringly with the patients whose stories are the warm heart of this timely and disturbing book."

—Daniel Perry, Past President, Coalition
for the Advancement of Medical Research, and
Executive Director, Alliance for Aging Research

"Herold's reporter-like style is effective as she sifts through various layers of the science and the social and religious controversies and provides an easily followed time frame of the major discoveries and events over the past decade in stem cell research, including the most recent revelation of scientific fraud in producing patient-specific embryonic stem cells. The issues with stem cell research are complex and Eve Herold is successful in presenting them in an easily understood fashion."

—John Gearhart, Johns Hopkins Medicine

"Eve Herold is an eyewitness to history. She chronicles the battle of patients and researchers to advance the greatest medical breakthrough of our lifetimes in this highly readable account of the rancorous public policy debate that has become the #1 wedge issue in American politics. As part of the chronicle of the world stem cell debate, Eve Herold presents the inside story of Woo Suk Hwang and the Korean cloning scandal, and supplies the shocking details about the misconduct that rocked all of medical science."

—Bernard Siegel, Executive Director, Genetics Policy Institute

"An outstanding science writer, Eve Herold makes the issues clear in a fascinatingly readable style. Engaging and clearly written, a must-have book to understand what's happening behind the scenes in a world where a political mistake could delay life and death research for decades. As someone who has been involved in stem cell politics essentially since it began, I learned a lot from this book!"

—Don C. Reed, Chairman, Californians for Cures

"Simple and to-the-point, this book gives a thorough explanation of what stem cell research is, why there's so much controversy surrounding it and ultimately, why our government must change course to allow it to flourish."

—June Walker, President of Hadassah

stem cell wars

Inside Stories from the Frontlines

Eve Herold

Foreword by
Dr. George Daley

STEM CELL WARS

First published in hardcover in 2006 by
PALGRAVE MACMILLAN™
175 Fifth Avenue, New York, N.Y. 10010 and
Houndmills, Basingstoke, Hampshire, England RG21 6XS.
Companies and representatives throughout the world.

PALGRAVE MACMILLAN is the global academic imprint of the Palgrave Macmillan division of St. Martin's Press, LLC and of Palgrave Macmillan Ltd. Macmillan® is a registered trademark in the United States, United Kingdom and other countries. Palgrave is a registered trademark in the European Union and other countries.

ISBN-13: 978–1–4039–8499–9 paperback
ISBN-10: 1–4039–8499–9 paperback

Library of Congress Cataloging-in-Publication Data

Herold, Eve.
    Stem cell wars : inside the frontlines / Eve Herold.
        p. cm.
    Includes bibliographical references.
    ISBN-13: 978–1–4039–7499–0 (hardcover)
    ISBN-10: 1–4039–7499–3 (hardcover)
    ISBN-13: 978–1–4039–8499–9 (paperback)
    ISBN-10: 1–4039–8499–9 (paperback)
        1. Stem cells—Miscellanea. 2. Stem cells—Popular works.
    3. Medicine—Miscellanea I. Title

QH588.S83H47 2006
616'.02774—dc22   2006043262

A catalogue record for this book is available from the British Library.

Design by Newgen Imaging Systems (P) Ltd., Chennai, India.

First PALGRAVE MACMILLAN paperback edition: September 2007

10 9 8 7 6 5 4 3 2 1

Printed in the United States of America.

To Amer, for your love and support.

contents

acknowledgments

This book is the culmination of over five years of interaction with a very diverse community of stem cell research scientists, bioethicists, journalists, and advocates on both sides of the issue, all of whom helped me to understand some aspect of this complex field. But most of all, it is the product of countless communications from patients and their loved ones, whether by phone, e-mail, or in person. I want to thank these anonymous individuals for sharing their stories with me. They made me aware of how urgent, how desperate, the need is for the development of treatments that go beyond the stark limitations of today's medicine. I could never thank each and every person who provided me with the impetus to write this book, but it is because of their questions and concerns that I felt compelled to try to provide at least a cursory understanding of this complex field. I hope to have cleared up at least a few of the misunderstandings on a subject where public awareness is so vitally important.

I also want to thank my agent, Ronald Goldfarb, for his steady encouragement, for believing in this project, and for finding a good home for it. Many thanks go to my editor at Palgrave Macmillan, Amanda Johnson, who took on the task of editing it with a passion and commitment that exceeded my expectations. Many thanks to all the people at Palgrave who have given the book so much care and attention.

Special thanks go to Bernard Siegel at the Genetics Policy Institute for his critical support and advice in the completion of the book. I also want to thank Frank Cocozzelli, Danny Heumann, and Susan Fajt for sharing their stories, which made such important contributions to the content. Art Caplan and Glyn Stacie also provided me with valuable insights in understanding much about the regulation and oversight of an issue that touches on some of the most sensitive ethical subjects in science. Thanks also go to Sean Tipton of the American Society of Reproductive Medicine and Dr. William Kearns and Sharon Covington at Shady Grove Fertility for contributing greatly to my understanding of the practice of assisted reproduction.

I owe my sincerest gratitude to the advocates who have worked so tirelessly to keep journalists like me informed of the steady stream of developments in every facet of the stem cell research field, especially Daniel Perry, former president of the Coalition for the Advancement of Medical Research, and Steve Meyer of the Stem Cell Action Network. Thanks also go out to the many research advocates toiling away in the state legislatures, whose stories have educated and inspired me.

Last but certainly not least, I want to thank my husband, Amer, for his seemingly infinite patience and support throughout the process of writing this book, and for his invaluable comments and critiques. Most of all, I want to thank him for believing that I could do it, which is perhaps the greatest gift one human being can give to another.

foreword

I recently cared for an adorable two-year-old child admitted to the Children's Hospital in Boston because of recurrent, severe infections. He had spent almost half his life living in the hospital, receiving a potent mix of intravenous antibiotics meant to sterilize him of virulent pathogens that would normally present no challenge to the immune system of an infant. But this child's immune cells lacked a single critical enzyme that left him unable to fight infection. His was a miserable and potentially fatal disease. One day he'd be playing cheerfully, an utterly cute and engaging toddler. But the next day he'd spike a high fever and become profoundly ill, irritable, and inconsolable. We'd launch heroic efforts to locate the source of his fever. We'd search for latent infection lurking in his lungs or belly or brain, sometimes even surgically removing a piece of tissue we thought might be the culprit. But typically we failed and were left to wonder whether his fever was the sign of an unusual bug that evaded even our strongest antibiotics, or whether we were unwittingly causing him misery with the toxic mix of drugs that poured into his veins every day.

Frustrated, we had no cure for this child's condition. Although kids with a variety of genetic immune deficiencies can be cured through bone marrow transplantation—a treatment that harnesses the regenerative power of blood stem cells—this child's siblings were not a tissue match, and in his condition, collective medical wisdom is pessimistic about the prospects for a bone marrow transplantation from a completely unrelated donor. Thus we were left with little more to offer this child than supportive, vigilant medical surveillance and treatment with antibiotics, hoping against the day when we could no longer subdue the infections adequately, and the child might die.

Although a single jarring case, this child represents countless others with incurable and potentially fatal illnesses who inspire me to pursue basic biomedical research. One cannot help but feel our society has a responsibility to provide for this child. By curing this child's condition we save a life, restore a family, and enrich a community. This book recounts the rise of a new scientific discipline that offers tremendous promise for conquering this child's disease. Stem cell biology allows me to envision a strategy to create a stem cell from this child's skin, repair its genetic defects, coax it to become blood-forming tissue, and provide that child with the curative bone marrow replacement therapy that is currently unavailable to him because he wasn't lucky enough in the genetic lottery to have siblings who share his tissue type. This book is also about the scientific, ethical, and political barriers to exploring this field to its fullest potential. I first met Eve Herold when she was working for the Stem Cell Research Foundation (SCRF), a fledgling philanthropic group devoted to funding research on stem cells. As the chair of the scientific review committee for SCRF, I was responsible for coordinating a committee of stem cell experts who critiqued grant proposals submitted to the SCRF. Both Eve and I were drawn to SCRF because we shared the conviction that basic biomedical research on stem cells would advance fundamental knowledge and lay the foundation for treating incurable and intractable diseases like diabetes, Parkinson's, spinal cord injury, heart failure, cancer, and countless others.

I last met Eve in Seoul, South Korea, where we were both visiting the laboratory of stem cell sensation Dr. Woo Suk Hwang, who had captured the world's attention with his claims of remarkable prowess creating customized, patient-specific lines of embryonic stem cells. I was in Korea on a scientific fact-finding mission, while Eve, working as director for Public Policy Research and Education for the nonprofit Genetics Policy Institute, was there to explore the regulatory and ethical framework under which Dr. Hwang's group practiced. We now know that both the scientific and ethical conduct of Dr. Hwang's work was tainted, leaving the field of stem cell research with a black eye. Such bruises heal, however, and the promise of stem cell research remains undiminished, even if now a more distant prospect. This book is Eve's vision of how and why this field must and will move on, a vision that I share.

Although stem cells have been studied for decades, only recently has the field of stem cell biology emerged as a distinct discipline of modern biomedicine. Stem cell biology unites scientists working in seemingly disparate fields around a common set of goals: to understand how stem cells can regenerate themselves without exhaustion (a property called "self-renewal"), and yet when called upon, differentiate into highly specialized cells of a complex tissue, thereby maintaining or restoring tissue and organ integrity. As you will learn in greater depth in this book, stem cells come in several varieties, each with very distinct properties and potential. All stem cells are not equal. The most extensively dissected and analyzed stem cell is the hematopoietic stem cell, the master cell for the entire blood system, alone responsible for the generation of red blood cells, white blood cells, platelets, and the immune system. Other well-studied stem cells exist for the skin, gut, muscle, and parts of the brain and nervous system, lungs, and liver, but not all tissues in an adult regenerate from stem cells. The stem cells in highly regenerative tissues are often referred to as "adult" or "somatic," although these names have limitations, chiefly because some stem cells behave like those in adult tissues even though they are isolated from non-adults, for example, a newborn's umbilical cord blood. The chief defining feature of adult, somatic stem cells is that they are restricted in the types of cells they can form, typically to the specific specialized cells of the very tissues in which the stem cells reside. Whether hematopoietic stem cells might be coaxed or tricked into contributing to tissues beyond the blood and immune system by clever bioengineering remains a subject of considerable interest, but as a matter of normal physiology, adult stem cells fulfill a restricted set of tissue maintenance and repair functions.

Standing in stark contrast to the tissue-restricted nature of adult somatic stem cells, embryonic stem cells, the master cells that can be extracted from early embryos, are naturally destined to become all of the cells of the body, a property called pluripotency. We can demonstrate the unique property or pluripotency for embryonic stem cells of the mouse by injecting them into a defective early mouse embryo that has four rather than the normal two sets of chromosomes. When left on its own, an embryo carrying four sets of chromosomes (a tetraploid) will form a placenta, but never a developing embryo. But if embryonic stem cells are injected into the tetraploid embryo, a normal mouse pup can be born whose body is formed entirely from the injected embryonic stem cells, clear proof of embryonic stem cell pluripotency. Stem cell biologists believe that human embryonic stem cells can likewise regenerate all of the cells of the human body. One of the major thrusts of stem cell biology is to understand how to coax embryonic stem cells to form a single cell type among the many hundreds of highly specialized cells that can be defined in the body. Studying how embryonic stem cells specialize provides an unprecedented and unique opportunity to observe human development and interrogate it experimentally in a petri dish. From such studies might emerge new insights into disease, new drugs, and even replacement tissues to replace and repair tissues ravaged by disease. The promise of such studies is compelling.

Embryonic stem cells can be isolated from two principle sources. First are infertility clinics, where literally hundreds of thousands of tiny embryos remain frozen. These miniscule clusters of between 6 and 200 cells, smaller than the dot on an i, might otherwise be discarded as medical waste by couples that have completed in-vitro fertilization and do not wish to have additional children. Some couples choose instead to donate their embryos to stem cell research, and indeed, hundreds of embryonic stem cell cultures, called lines, have been made by scientists throughout the world, thereby providing invaluable tools for research. These are generic stem cells that can be used to ask basic questions about how embryonic stem cells behave. A second source of embryonic stem cells is potentially even more valuable: embryonic stem cells generated from a specific patient. Exploiting a method called nuclear transfer, which has worked in the mouse but has yet to succeed in humans, scientists hope to create customized patient-specific embryonic stem cells by inserting a patient's skin cells into the milieu of an egg whose own DNA has been removed. By a remarkable process deemed nuclear reprogramming, the skin cell reverts to an embryonic state and forms a cluster of cells resembling a normal embryo that has little or no reproductive potential but can yield embryonic stem cells. These embryonic stem cells carry all of the genetic baggage that contributed to that patient's disease. These disease-specific cells provide stem cell scientists with a new tool for medical research, and a potential source of replacement cells that are in essence the patient's own, and thus not subject to immune rejection like the tissues or organs of an unrelated donor. Over four years ago my lab published a scientific paper demonstrating that we could restore immune function in a mouse model of immune deficiency using precisely this procedure. I want to extend that work to generate stem cells for my two-year-old patient with immune deficiency, repair the genetic defect, differentiate the cells into hematopoietic stem cells, and transplant that child with his own normal tissue to repair his immune system. However, despite our success in mice, I have not yet been able to begin comparable experiments with my patient's cells. The social and political forces behind this frustrating state of affairs is the subject of this book.

Almost five years ago, President Bush announced a policy governing the provision of federal funds for embryonic stem cell research. From one perspective, President Bush endorsed the field by enabling funding of research on preexisting cultures of human embryonic stem cells. But the political tightrope he attempted to walk that day has grown slack, as folks on both sides of the issue remain unsatisfied: the opponents because the research has been allowed to proceed; the supporters because the compromise is so restrictive that it hinders robust growth in the field. We scientists are an energized lot of persistent and meticulous truth-seekers, who thrive on working with the latest equipment and the most up-to-date tools. When constrained by inadequate resources—too little funding, outmoded machinery, or poor access to key research materials—scientific progress is severely curtailed and scientific morale is dealt a blow. And yet this is precisely the current state of affairs for promising areas of stem cell research. If stem cell biologists want to obtain federal funds for their research, they must agree to use only a small set of less than two dozen embryonic stem cell lines that were made prior to the 2001 date of the president's policy announcement. Many of these lines are now outmoded or in disrepair. Since 2001, hundreds of new lines have been created throughout the world, many with advantageous features for medical research and treatment. And for those of us looking to practice nuclear transfer to generate patient-specific embryonic stem cell lines, there are no prospects for federal funding, and we are left to seek private philanthropy.

Distilled to its very essence, the controversy surrounding stem cell research pits the scientists who believe that embryonic stem cells and, in particular, customized patient-specific embryonic stem cells offer great promise for biomedical research, against those who believe that the human embryo is an inviolable being that should be accorded full status as a members of society, and, as such, protected from harm. The isolation of embryonic stem cells from a human embryo destroys that embryo. Those who believe that the embryo is a person view extracting stem cells as murder, and no appeal to the benefits of stem cell research will justify its practice.

I do not find the arguments defending the rights of embryos compelling enough to warrant prohibitions or even significant restrictions on embryonic stem cell research. Over the last decade I have found myself devoting countless hours to justifying stem cell biology, at the expense of progress in my own research. I rationalize these diversions because a scientist must also be an educator. I and my colleagues in the stem cell field have been called upon with an unprecedented frequency to teach the principles of stem cell biology to curious members of the media, to various legislative bodies at the local, state, and national levels, and, of course, to the public, through community lectures, coffee shop socials, and adult education events at churches and synagogues. The effort is paying off. Opinion polls have reflected a steady increase in public support for all forms of stem cell research. The politicians cannot be far behind.

What most of the public is reflecting is a moral perspective that accords the human embryo a unique and weighty status, but does not view the embryo as a person. The prevailing public sentiment is that patients suffering from disease make more immediate and compelling claims on our society to ensure their well-being than do embryos in a freezer or a petri dish. There are many reasons why embryos are not thought of as people by the vast majority of the public. Despite arguments that conception represents the beginning of a new and unique life, and that embryos should be considered human beings, most people's moral intuition, and indeed the theological perspective of major world religions like Islam and Judaism, see the acquisition of moral status not as a clear bright line beginning at conception, but rather as a special status acquired some time later in human development, especially as we emerge as sentient and biologically independent beings.

Biology itself does not support the notion of a "moment" of conception. In fact, conception is a complex process that proceeds over many hours, and although a new genome is formed when the egg and sperm pro-nuclei fuse to become the single-celled human zygote, a unique biological individual is not apparent until later in human development. For at least the first two weeks of gestation, the early embryo can split, forming twins, triplets, and, rarely, quadruplets or more. And pairs of fertilized eggs that might otherwise generate fraternal twins can aggregate in the womb to form a single normal individual that carries the genetic complement of two distinct conceptions, a phenomenon called tetragametic chimerism. It is hard to consider the early embryo a person if it is divisible, because individuality and uniqueness of spirit are intimately tied to our notions of personhood. Finally, should one consider a person to be formed when nuclear transfer is used to generate a patient-specific embryonic stem cell? Some will argue that a prohibition against using federal funds in embryonic stem cell research is justified because it is wrong to force taxpayers who have strong moral objections to financially support the science. However, there are many subjects that do not garner moral consensus and yet are fully supported in the federal budgets, like research on animals that a vocal minority of animal-right's activists oppose. We should appreciate that the policy to restrict federal funding for embryonic stem cell research is a political decision imposed by politicians who wish to advocate the rights of embryos.

The current debate over embryonic stem cell research and nuclear transfer has parallels with the debate that followed the birth of the first test tube baby in England in 1978. There was a similar though less long-lived controversy about the propriety of in-vitro fertilization (IVF). IVF was considered by some an abomination—a grave threat to humanity, destined to usher in a future of mechanized, dehumanized human reproduction. Today, IVF is a routine part of medical practice and responsible for fulfilling the hopes of tens of thousands of couples a year who are able to give birth to their own children. I believe that some twenty to thirty years from now, when stem cell science has proven itself a powerful force in biomedicine, and cell-based therapies are the standard of care for a range of diseases, we will reflect on the stem cell debate and see it in its historical context. It will come to represent just one example of an ever-accelerating series of challenges to society posed by rapid technological change. Such change is unsettling. It threatens our traditions and compels us to make new choices in deeply personal arenas like reproduction. But societal change compelled by technological innovation is an inevitable feature of a curious and inventive people. We as a society must confront technological change with intelligent and reasoned debate and make thoughtful choices. I believe that we can pursue stem cell research in a responsible manner, so that its benefits will outweigh concerns for a dehumanizing effect on society. Indeed, I believe that the current debate over stem cell research will play a central role in our society's coming to terms with the profound influence that biomedical research will have on our future.

George Q. Daley, MD, PhD
Children's Hospital, Boston
March 2006

chapter one

the field of battle

Three passions, simple but overwhelmingly strong, have governed my life: the longing for love, the search for knowledge, and unbearable pity for the suffering of mankind.

—Bertrand Russell

Doctors are men who prescribe medicines of which they know little, to cure diseases of which they know less, in human beings of whom they know nothing.

—Voltaire

Frank Cocozzelli's life should have turned out differently. Born in Brooklyn, New York to devout Italian Catholic parents, Frank was an exceptionally bright boy. Unlike most of the boys he went to school with, he was more interested in intellectual heavy lifting than in sports. Graduating in 1982 from Queens College, he continued on and became part of the first class to attend Queens' new law school, pursuing his goal of becoming a lawyer.

While in law school, Frank decided to try to overcome what he saw as his lack of athletic ability by joining a gym, where he started lifting weights.

The weight-lifting was challenging at first, but by steadily applying himself, he was soon bench-pressing 140 to 150 pounds. Then, one day, he noticed that instead of gaining strength in his upper body, he was actually losing it. Inexplicably, week after week, he could lift less and less weight. It made no sense for a young guy in his twenties to grow weaker instead of stronger.

Around this time, he decided to make a little spending money by taking a part-time job as a "gofer" for an attorney in Queens. One of his duties was to walk to the bank, which was about half a mile away, and make deposits for the law firm. He had no trouble getting to the bank, but on his way back to the office, he would suddenly be seized by severe pain in his legs. It seemed to be brought on by exertion. He was also developing a noticeable limp. Concerned and completely baffled by this new development, Frank made an appointment to see a neurologist in Long Island who happened to be treating his grandfather for Parkinson's disease.

The neurologist conducted an initial examination, and told him that he thought the problem was probably a pinched nerve, caused by a dislocated vertebra, a relatively common condition that can be corrected by surgery. Frank went home relieved, but a few days later, the results of one of the tests came back, and the news wasn't just bad, it was devastating.

The test that rocked Frank's world is called an electromyography. It involves placing a needle into the skin at the top of the thigh, and another one at the base of the foot, then shooting an electric current through the muscles to see how they respond to stimulation. And Frank's muscles showed signs not just of weakness, but of pathological atrophy. When another test came back showing abnormal levels of an enzyme called creatinine kinase in his blood, the worst was confirmed: The source of Frank's problems was muscular dystrophy.

Muscular dystrophy, which is passed down on one of the mother's X chromosomes, occasionally strikes females but shows up far more often in males. It's more common than one might think—approximately 30 out of every 100,000 male babies are born with the genetic defect for some form of muscular dystrophy. Some of them are lucky enough to develop only a mild form of the disease and don't become wheelchair-bound. But most do, and those who have the most common forms of the disease—Duchenne's and Becker's muscular dystrophies—suffer progressive muscular atrophy so severe that they often die before the age of 20. Frank's variant of the disease usually shows up in early adulthood.

The root of the progressive muscle weakness is that the muscles are unable to make a key protein called dystrophin. Because this protein is critical for muscle cells to maintain their structure, the muscle fibers first enlarge, then progressively die off and are replaced by fat and other useless tissues. People with muscular dystrophy grow weaker and weaker, gradually becoming so weak that they can't move. At some point, they are usually confined to a wheelchair, then bedridden, in a slow descent into greater degrees of helplessness. All the while their minds remain intact. Like the victims of amyotrophic lateral sclerosis (ALS), or Lou Gehrig's disease, they become prisoners in a body that needs 24-hour-a-day care and help with every basic function. Along the way, because of the patient's inability to move, he develops painful muscle contractions and severe osteoporosis, which leads to brittle, easily fractured bones. Pneumonia and other infections can easily overwhelm his compromised immune system, and these are often the immediate cause of death. In very aggressive forms of muscular dystrophy, the heart muscle weakens until it can no longer pump.

Doctors are helpless in the face of this devastating condition. They can prescribe physical therapy to try to slow the degeneration of muscles and to ease some of the pain caused by permanent muscle contraction. But there's no way to halt the disease, which runs its inexorable course and cuts the victim's life span short by several decades. One of the most difficult things for patients and their loved ones to cope with is that doctors can't even tell them how long they will live or at what level of disability. Every case is individualized, and every morning that a muscular dystrophy patient wakes up is fraught with uncertainty about what tiny but infinitely precious ability will be lost that day.

Today the sheer force of Frank's mind belies the weakness of his body. His memory is encyclopedic, enabling him to rattle off names and dates and long-ago events with a fluency that is nothing short of amazing. He is considered one of the "lucky ones," having one of the rarer forms of the disease, a variation called limb-girdle muscular dystrophy, or LGMD.

He has made it to the age of 47, although he has spent the last few of those years in a wheelchair. He is losing more and more of the movement in his arms and requires around-the-clock assistance, which is provided by his wife and his father.

"I have to be lifted in a sling to go to the bathroom," he says matter-of-factly. "It's very difficult for my wife and father to lift me. I can lift my arms a little, with help. And I can still feed myself, although I'm told it looks disgusting. I have to hunch over the plate and push the food up to my mouth, and it goes everywhere. People don't like to watch it." Frank is sensitive about eating in public, afraid that he might offend people in restaurants. When asked how he copes with the progression of his illness, he says, "I try not to think about what tomorrow brings." He focuses instead on fighting the small battles, of using what little movement he has left so that he can retain it as long as possible.

Frank would be the first to tell you he is not a victim. He married his college sweetheart, who knew about the muscular dystrophy, and together they had two children. He still practices law, and goes to his office in Garden City, New York, two or three times a week. He is also a stem cell research activist, and three years ago, he and a partner founded the Committee for the Advancement of Stem Cell Research, a political action group. A few years ago, he learned about the possibility of stem cell treatments curing muscular dystrophy.

In 1998, Frank traveled to Miami to see a new doctor, an expert on paralysis at the University of Miami. After years of being given no hope, this doctor explained to him how, if the research succeeded, treatments derived from embryonic stem cells could possibly replace dead muscle tissue by transplanting the precursors of normal muscle cells into his body. If the treatment worked, the stem cells would divide into healthy muscle cells able to make the dystrophin he needs to rebuild his body and recover his strength. "We're really hopeful about this one," the doctor told him. Although the research was at the very early stages, it was the first glimmer of hope Frank had been given in the 20 years since his diagnosis.

At the time, President Clinton was still in office, and, in light of the 1998 isolation of human embryonic stem cells, plus a string of incredible successes in animal studies, Clinton had put together a commission to review the existing federal policy of withholding funding for research using embryos. Rules that had been in effect since 1993 forbade the U.S. government from funding research that involved the destruction of human embryos, but evidence was mounting through animal studies that embryonic stem cells held extraordinary promise for the treatment of a huge range of diseases. "I'll never forget a certain day in 2000," Frank recalls. "My wife was dressing me for court, and we had *The Today Show* on. They said that Clinton was reversing the ban on embryonic stem cell research. I got so hopeful, it was all I could think about that day. I thought, 'Maybe this is it. Maybe there really is going to be a cure.' "

But Frank's hopes were not to be realized, at least not that year. The Clinton presidency was drawing to a close and the 2000 presidential race was on. Al Gore had clearly stated that he was in favor of allowing the National Institutes of Health to fund embryonic stem cell research, but candidate George W. Bush had already made public his opposition. Like countless other patients with otherwise incurable diseases, the days when the 2000 election was still undecided were an emotional roller coaster for Frank. Then, the Supreme Court made its decision in favor of Bush. "I'll never forget the night I heard their decision," Frank says. "I could still drive then, with an assistive device, and I was sitting in my car listening to the radio. They made the announcement, and I just sat in my car and cried."

One of the first actions that George Bush took after becoming president was to put the brakes on the Clinton proposal regarding embryonic stem cell research before it could be implemented. In the months that followed, crushing disappointment for people like Frank turned to anger. He was soon to realize that Bush's 2001 "compromise," to allow federal funding of research on existing embryonic stem cell lines (or batches of cells derived from a single embryo), was a swindle for those who were waiting for cures. That realization, however, coalesced over a matter of weeks, when it was discovered that many of the cell lines that the president said were approved for federal funding hadn't been properly characterized or had died. Others turned out not to be embryonic stem cells at all, while others were located in countries that forbid the export of biological materials, such as India. It became clear that of the 64 cell lines initially claimed by President Bush, only 22 were actually both viable and available.

Even as some journalists were lauding the president for "opening the door" to stem cell research, scientists and patients were gradually learning the true nature of the federal limitations. Those limitations meant that U.S. government support for embryonic stem cell research was little more than a token. What the administration had done was shore up a massive dam against funding for embryonic stem cell research, while allowing a tiny trickle of research dollars to leak through. By the time scientists, patients, and advocates realized the true scope of the decision, President Bush had already scored a major public relations victory. In the four years since that decision was announced, Bush has said repeatedly that he will not allow any loosening of the federal restrictions on embryonic stem cell research.

Since 2001, Frank's frustration with the Bush administration has only intensified. He feels especially angry that Bush and his evangelical constituents seem to imply that being anti-embryonic stem cell research is the only legitimate religious view. Still a devoutly religious Catholic himself, he decries the fact that President Bush " . . . ignores the pro-research views of most American Catholics," obliging instead the ultra-conservative "Opus Dei" segment of Catholics, represented by politicians such as senators Sam Brownback (R-KS) and Rick Santorum (R-PA). "Furthermore," he notes, "on human embryonic stem cell research, he has selectively adopted the Vatican's position, but he chooses to ignore the pope in his opposition to the Iraq war, a situation that just raises his frustration level even higher."

As a random victim of illness, Frank Cocozzelli is far from alone. There are millions of Americans like him who search desperately for the help that, all too often, medicine cannot give them. Many diseases are part of the cruel randomness of the genetic lottery. Disease-related genetic mutations can lie silent for decades, until we reach a certain age or experience some environmental trigger that sets the disease process into motion. Others are evident from birth.

In late 2004, I got a phone call from the grandfather of a newborn baby girl. I'll call him Alan. Alan had been up, around the clock, for days, ever since his granddaughter was born two weeks previously. He had been searching the Internet desperately for some type of therapy that might help her. He found the website of the Stem Cell Research Foundation and called our toll-free number.

He told me that his granddaughter was a beautiful baby who looked normal in every way, but she was born with a condition called triple X syndrome. This meant that instead of having two X chromosomes, which is normal for females, her cells had three. Strangely enough, babies born with this condition develop normally in every way, with the tragic exception of their brains, which are subject to devastating seizures. I could hear exhaustion, fear, hope, and urgency in his voice as he described how this baby's tiny body was being racked with 15 to 20 violent seizures a day. "She will seem fine one minute, and then her eyes roll back and she just curls up," he said. Even worse, the doctor had told the shocked and distraught family that the electrochemical haywire searing through her brain every time she had one of these seizures was rapidly destroying her hope of a normal life.

"Every seizure she has is doing more damage to her brain," Alan said. "The doctor said what it's doing to her brain is like a computer shutting down 15 to 20 times a day. After a while it fries the circuit. If we could do something now to stop the seizures, she would be less damaged. But if this goes on, we've been told she won't develop beyond the mental age of a four-year-old, at the most." Like so many other people, Alan was hoping that stem cells would offer some kind of treatment, even if it was experimental. An experimental treatment, even with a totally uncertain outcome, still sounded better than what the doctors were telling this family.

I asked him what type of treatments the pediatricians were giving her. He said they were giving the tiny newborn some powerful anti-seizure medication that turned her into "a zombie." And still, the seizures returned. "It just comes over her, and it's violent. She struggles with it until she's exhausted. And then after a while she comes back," he said. "But I know that one day she's not going to come back."

It wasn't easy telling this desperate grandfather, who was ready to do anything humanly possible to help her, that his daughter's first baby was in a race against time that she had little to no chance of winning. If it were simply a matter of replacing damaged brain tissue with healthy new cells, then stem cell scientists are now closing in on a way to do that. But in addition to repairing damaged brain tissue, what this baby needed was a way to repair the genetic glitch in every cell of her body, and that is something that is an untold number years into the future. Many scientists believe that gene-based cures, perhaps combined with stem cells as the carriers of corrected genes, could someday cure syndromes like hers, but the best time to intervene in such a case would be long before birth, perhaps even at the embryonic stage.

What happened to Alan's granddaughter happened very early in her development, when the first cells with the genetic mistake copied themselves over and over as they divided and formed her body. But the only way to have detected the condition early enough to prevent it would have been if Alan's granddaughter had been conceived in vitro—in the lab— and if her embryo had been examined within a few days of its existence. And, problematically for some, scientists will never be able to diagnose and correct such mistakes without first doing research on human embryos.

There's no doubt that medical science made some spectacular strides in the twentieth century. In fact, more effective treatments and cures were discovered during the last century than in all of prior human history. New drugs, diagnostic techniques like x-rays, CT scans, and MRIs, plus the ability to do ever more complicated procedures, such as heart bypass surgery, have benefited millions of people whose lives would otherwise have been cut short. If anyone doubts the march of medical progress, the most dramatic testament is the fact that, between 1900 and 1999, the average life span for Americans increased from 47 years to 77, and every few years that number is adjusted upward.[1]

We're quick to credit modern technology and space-age research advances for the increased longevity that has occurred throughout the developed world. But the leap in better health actually owes less to advanced technology and technologically based research than it does to the decidedly humble issue of hygiene. Before modern sewage disposal and water purification methods, water-borne infectious diseases like malaria were the scourges of mankind, raging through towns and villages, leaving enormous death tolls in their wakes. For centuries, the slums of Europe were periodically ravaged by all kinds of plagues carried by rats, fleas, and other vermin that thrive in unsanitary conditions. Running water, clean drinking water, better trash and sewage disposal, and public education about simple hygiene have saved millions of lives.

The other dramatic improvement came in the form of treating and preventing infectious diseases like tuberculosis, polio, measles, and diphtheria, which used to kill huge numbers of babies, children, and young adults in the prime of their lives. These diseases have been all but wiped out in the industrialized world because of the development of vaccines. And the discovery of antibiotics essentially disarmed another common killer—pneumonia—which claimed the lives of people of all ages. Today pneumonia is generally lethal only to the very old and sick, or those with compromised immune systems.

But twentieth-century progress has come with a price. Many more people throughout the world are living with incurable genetic diseases and with the chronic, degenerative diseases that become more prevalent after the age of 40. The modern dilemma of more or less "successfully" treating so many diseases without actually curing them has become the hallmark of later life. For example, more people survive heart attacks only to suffer the slow deterioration of heart failure. Thanks to clot-busting drugs, more people survive strokes only to live with chronic disabilities such as memory and speech loss and paralysis. And more people are living long enough with diabetes to suffer its more sinister complications, such as heart disease, severe visual impairment, and kidney failure. The list goes on and on, but perhaps one of the most tragic ironies of living longer has been the rapid increase in the incidence of Alzheimer's disease.

Alzheimer's disease has become a symbol for today's double-edged sword of aging. The risk of Alzheimer's, while not synonymous with aging, climbs so quickly after the age of 65 that, with America's aging population, it threatens to become an epidemic over the next few decades. At age 65, our chance of having Alzheimer's is approximately 10 percent; by age 85, you have nearly a 50/50 chance of having it.[2] Just looking at the country's aging demographics is enough to envision a future with armies of frail older people who need constant care. Some social scientists even depict a future in which practically everyone will either have Alzheimer's or be caring for someone who does.

But Alzheimer's is only one of many age-related diseases that is striking with far greater frequency than ever before. There's the ever-rising risk of cancer, which goes up every year after the age of 40. According to the National Cancer Institute (the cancer research arm of the National Institutes of Health), from birth to age 39, American males have a 1-in-71 chance of developing some type of cancer, while for young females, the odds are 1-in-51. But from age 60 to age 79, one in *three* males will develop cancer and one in *four* females will.[3] It's true that better treatments, including more effective and less toxic chemotherapy drugs, are helping more people live for years after diagnosis. Doctors now regard some cancers as chronic conditions that can be "managed," at least for a considerable period of time. However, for the first time, and partly because of the aging of the population, cancer rivals heart disease and obesity as one of the leading causes of death.

In spite of the modern medical "miracles" that have almost become a cliché, more than one-third of Americans are now living with a chronic, degenerative health condition. This includes 60 million Americans with some form of cardiovascular disease (including coronary heart disease and stroke), 16 million diabetics, over 8 million cancer patients, 30 million with an autoimmune disease, 10 million with osteoporosis, 4 million with Alzheimer's disease, and over one and one-half million with Parkinson's disease. Add to that the victims of spinal cord injury, severe burns and other serious injuries, osteoarthritis, multiple sclerosis, muscular dystrophy, and chronic kidney, lung, and liver disease, and the number of chronically ill Americans easily surpasses 128 million.

The litany of today's common diseases suggests that medical science has an almost overwhelming task in finding cures for them. Based on the historical rate of biomedical research and the methodical search for cures, it's easy to imagine that the next century could be filled up with finding cures for them one by one. But that may not be the case, because all of the above conditions have certain features in common.

First of all, they are not, as far as we know, caused by contagious pathogens. In other words, they can't be cured by antibiotics, as they would if they were caused by bacteria, and they are not caused by viruses, which would be amenable to prevention by vaccine. They are a class of diseases that represent the next frontier in medicine—incurable, degenerative conditions of unknown or complex origin, that cause some sort of malfunction at the cellular level. In other words, these are *cell-based diseases*. They cause the disruption of one or more vital cellular processes, which over time leads to cell degeneration and cell death.

Frank Cocozzelli's problem involves the failure of a vital function of his muscle cells. For those with Parkinson's disease, the problem occurs in brain cells that fail to make a critical neurotransmitter called dopamine, and in heart failure, cardiac cells enlarge and lose their ability to "beat," impairing the heart muscle's ability to pump blood. In multiple sclerosis, the cells that surround and insulate nerve fibers die, and the body cannot replace them. Cells are far more than the bricks and mortar of the body—they perform countless specialized functions, many of which haven't even been named yet. They can be compared to little engines or factories, converting food into sugar and sugar into energy, assembling chemicals and hormones, sending and receiving signals from other cells, eliminating wastes, and even conducting electricity. You name it, and if the body needs it, there is some specialized cell dedicated to producing and metabolizing it.

But cells can stop working or make mistakes when they copy themselves (or divide), and these mistakes slowly accumulate as we age. It is a little-known fact that the DNA in each of our cells is not immutable—fixed for as long as we live—but actually suffers wear and tear throughout life. Ubiquitous background radiation, which is a natural phenomenon, and other environmental corrosives steadily chip away at our cells, including the genes that tell them what to do. So do sunlight and toxic chemicals that we ingest. And then there's the inevitable process of living itself. When our cells convert oxygen to energy (a process that must take place continuously in order to sustain life), damaging byproducts called oxygen free radicals are constantly being created. These free radicals (unstable atoms with one too many electrons) steadily chip away at cellular genetic material, and tiny glitches creep into the cell's genetic program, accumulating until some critical process is disrupted. Cellular dysfunction leads over time to tissue dysfunction and, at a certain tipping point, organ failure and eventually, death. This is the case for a huge array of diseases, in which a specific cell type becomes sick and, over time, dies.

Injuries and traumatic events can also cause massive cell death, as in the case of head traumas and spinal cord injuries, in which the nerve cells critical for thinking or for movement cease to function or die. A stroke or a heart attack, by blocking the blood supply to the brain or heart muscle, causes the death of large numbers of specialized cells, which the body cannot replace in sufficient numbers to restore the organ's vital functions. Just for the sake of discussion, let's assume that every human being has the capacity to make and use 100 trillion cells over the course of a lifetime. Eventually, every one of those 100 trillion cells will die. They won't die all at once—some cells will die more quickly while others are happily chugging along, but even in some perfect world where there was no disease, eventually every last one of them would blink out.

Whether we're talking about the brain, the heart, the kidneys, liver, pancreas, cartilage, skin, or bone, from conception to birth to old age and death, the human body can only make or replace a finite number of these cells. When even one of the body's approximately 200 cell types dies, some infinitesimal part of a necessary function is lost. This cold, hard fact has long been regarded as a hard-wired biological limit, a final obstacle that medicine has been utterly unable to overcome in any way other than through organ transplantation.

If "round one" for modern medicine was the eradication of infectious diseases, then we are close to claiming victory, at least in the industrialized world (with the major exception of AIDS). Now that most of us live long enough to develop a chronic, degenerative disease, "round two" will no doubt consist of the effort to conquer disease at the cellular level. And up until the last few years, we were nowhere close to being able to do that.

Perhaps one of the most widespread myths in American society today is the myth of the triumph of modern medicine. We journalists love to report on amazing medical breakthroughs, but in fact, the perceptions that many of us have regarding the medical treatments that are available, and the effectiveness of those treatments, are far out of step with reality. Some of the awe we have for medicine is undoubtedly deserved, because of specific life-saving breakthroughs like heart bypass surgery and organ transplant. But in general, the mass media, which is most people's source of information about progress in medical research, has at best a mixed record in medical reporting. Mainstream journalists, struggling to follow the fast-moving world of research, often misunderstand biomedical information, at times hailing the results of one experimental study as the harbinger of a "miracle cure" while ignoring more important and truly meaningful developments.

Few journalists are aware of how painfully slow and difficult it is to obtain and interpret research results, then duplicate and verify those results, then translate them into clinical treatments. The frequent over-reporting of early experimental results in the lab is one reason why scientists tend to be extremely conservative in the conclusions they draw about the possible applications of their work. If you ask almost any biomedical researcher today, you will quickly learn that medicine is in its infancy.

After centuries of merely managing (and mismanaging) disease, only recently has science begun to make headway in its quest to actually *cure*. And still the cures seem to stand as lonely monoliths against the tidal wave of illnesses, injuries, and birth defects that afflict the human race. Modern molecular biology is just beginning to give scientists the knowledge they need for the intelligent design of drugs, as opposed to the hit-or-miss method that has been used up until this day. Gene-based therapies still face technical obstacles that must be overcome before cures based on them become a reality. Recently, the human genome was decoded, meaning that the sequencing of chemical pairs on chromosomes has been identified. But now the real work begins—of finding out what part each gene plays in the complex orchestration of the body's development and its countless processes throughout life.

Organ transplantation has offered new life to thousands of sick patients, but there is a chronic and critical shortage of organs to transplant. Out of the 89,000 Americans who are currently on organ transplant waiting lists, more than half will die waiting. And for those lucky enough to receive a new heart, kidney or pancreas, the specter of rejection looms darkly over them for the rest of their lives.

Even the widely-touted "cures" of cancer can entail enormous collateral damage to the body. Radiation and chemotherapy can ravage healthy tissues, causing as much or more damage than the cancer itself. More and more, we are forced to ask ourselves, at what price do we want life? This question grows especially acute toward the end of life, when the continuation of vital functions merely prolongs suffering. By methodically and aggressively addressing the individual functions that sustain biological life, medicine can (and often does) keep our bodies hanging on after our minds are long gone. But a twilight existence that is dependent on machines is not what we want from medicine. That realization was experienced by a huge majority of Americans as they witnessed the macabre struggle surrounding the April 2005 death of Terri Schiavo. Despite the views of those who fought to keep Schiavo's feeding tube in place, most of us believe that medicine has not succeeded when a mere minimum of biological function is sustained. What we want and need is to be cured, at least to the point where we feel that our lives can have some meaning.

Just imagine what it would be like if doctors could actually cure a huge number of the most common diseases, rather than just managing them, by getting around the limits of our finite allotment of cells. Suppose that whatever cells were being destroyed by a disease could be replaced by healthy, functioning ones. What if a diabetic, for example, could have her pancreas replenished with a fresh supply of insulin-secreting islet cells that would become a permanent part of her body? Or if a stroke victim could have healthy new brain cells to fill in his brain's "dead" regions of oxygen-starved tissue, leading to a complete recovery? If a child with cystic fibrosis could have his lungs "seeded" with living stem cells that would repopulate his airways with healthy lung tissue that would be his for life? If a Parkinson's patient could have his symptoms completely reversed by a restoration of dopamine-producing brain cells? Or if a person dying of kidney disease could have her failing organs rebuilt from the inside out by healthy, functioning kidney cells that would copy themselves over and over. In the foreseeable future, stem cell research may allow us to do these things and much, much more.

Using stem cells, scientists have already been able to produce several specialized cell types in the lab. These cells, when transplanted into sick organs, could permanently cure a huge range of diseases by becoming a living, permanent part of the patient's body. Unlike drugs, which work only fleetingly, until their chemical compounds are broken down, cellular transplants can provide the tiny factories—the cells—that work around the clock, producing the chemicals, hormones, and other molecules that our bodies need in order to function. If cellular transplants become as effective as scientists think they will, the implications for medicine are staggering. A huge number of currently incurable diseases could be wiped out (or staved off in an individual's life by decades), and the human life span could be dramatically extended. And not just the life span, but just as importantly, the *health span* of humans, meaning that old age and its infirmities could be significantly postponed and youth and middle age extended by decades.

Stem cell research, especially research using human embryonic stem cells, is only a few years old, but the field has already yielded swift and surprising results in animals and in the laboratory. It is considered by the world's brightest scientific minds to be a true revolution in the way we view and treat disease. We are not that far from the day when heart muscle that has been damaged or killed by a heart attack could be revived through an infusion of healthy cardiac cells, adding healthy decades to a patient's life. Human embryonic stem cells have already been used to create dopamine-producing neurons (the cells that are lost to Parkinson's disease) and motor neurons, the cells that could cure ALS or reverse paralysis in a stroke victim. Rats that have had their hindquarters paralyzed due to spinal cord injury have been able to walk again due to an injection of human embryonic stem cells. The cells traveled to the sites of damage, differentiated into the proper cell types and even healed injured nerve cells that hadn't died but were damaged and in distress. Human stem cells have also given rise in the lab to living retinal cells—some of the body's most precious cells for their sight-giving ability. Scientists at Harvard now believe they have actually reversed blindness in mice through the transplantation of retinal stem cells into their damaged retinas. Scientists at Duke University and elsewhere have used stem cells to grow new skin, bone, and cartilage—developments that could be a true godsend for severe burn victims, people who have sustained serious injuries, and victims of bone cancer, osteoporosis, and osteoarthritis. In fact, age-related arthritis may soon be relegated to history because doctors may be able to stimulate our bodies to regrow their own supply of cartilage. As over-the-top as this sounds, these possibilities are amply supported by science, as we'll see in chapter three.

Eventually, as the science of harnessing and manipulating stem cells progresses, thousands of people could be taken off of transplant lists. Organs and tissues that have been ravaged by cancer could be repaired from the inside out, or, failing that, an entire organ could be grown. This is not science fiction—scientists have already grown functioning human bladders through a combination of stem cell technology (using the patient's own stem cells) and tissue engineering.

Scientists caution, however, that the human body and the myriad of diseases it is subject to is intricately complex and involves countless variables introduced through our lifestyle, habits, and interaction with the environment. As tantalizing as the research has become, at the very dawn of this new era in medicine, there are daunting scientific hurdles that must be overcome for the dream of widespread stem cell treatments to become a reality. But if even half of what scientists hope for materializes, it would radically change the state of human health. Millions of people could be spared the pain and suffering, the crushing disabilities, and the countless indignities of prolonged illness. Whole societies would be dramatically altered by the increased health, longevity, and productivity of their people. This is no small consideration on a planet where advanced aging is becoming a worldwide phenomenon.

However, some of the most important stem cell research is not proceeding at the pace one might expect given its tremendous potential. In fact, in the United States, the most biomedically advanced country in the world, some types of stem cell research are in danger of outright criminalization. Even though a majority of Americans are strongly in favor of this revolutionary science, it is being fought vigorously—vociferously even—by powerful conservative think tanks and activist organizations of the religious right. These groups equate embryonic stem cell research with murder, and they are throwing as much skill, audacity, and public relations savvy—not to mention hundreds of millions of dollars—at the issue as they can. They've launched anti-embryonic stem cell research campaigns with staggering swiftness, considering that human embryonic stem cells were only isolated for the first time in 1998.

The fact that these groups are able to organize large constituencies of grassroots advocates, to saturate the media with their message, and to hamstring governmental bodies should come as no surprise. Many of them have had several decades of practice in the fight to ban abortion. Many believe that if they can ban embryonic stem cell research, they will be one giant step closer to criminalizing abortion. As inaccurate as the association is, they have managed to deeply entangle the issue of embryonic stem cell research in millions of people's minds with the practice of abortion. As I'll explain in chapter four, some organizations see the banning of embryonic stem cell research as instrumental to their overarching strategy to institute antiabortion legislation.

With George W. Bush in the White House (thanks in large part to the support of evangelicals and other religious conservatives), Republicans and Democrats alike tremble with the fear of becoming the next target of right-wing think tanks and activist organizations like the Family Research Council and James Dobson's Focus on the Family. In their zeal to ban embryonic stem cell research, these groups have initiated fierce legislative battles on the national level and in virtually every state. They have hundreds of conservative politicians fighting for their agenda in state legislatures, the U.S. Congress, and the Senate. This has meant that in federal legislation as well as in almost every state legislature, for every law protecting or supporting embryonic stem cell research, there is a countervailing piece of legislation severely criminalizing it. For every step toward the establishment of adequate funding for embryonic stem cell research, there is at least one step backward. While scientists fight for the right to study microscopic cells in lab dishes, lawmakers are fighting fierce battles to turn these scientists, plus patients, doctors, nurses, and even the relatives who assist their loved ones in getting treatments, into felony-committing criminals. For almost every check there has been a checkmate.

Opponents of stem cell research have managed to carry on a massive campaign of misinformation, to the extent that millions of Americans are either deeply confused about what the research entails or actually believe that mainstream scientists routinely dismember and kill human babies for research. These extremists create fictions that are stunning in their outrageousness, yet these tactics have gotten them so far that they've managed to hamstring the nation's universities, the National Institutes of Health (which administers 95 percent of the nation's research dollars through government grants), and most of the scientific community from conducting what many believe to be the most promising research of our time. For what these anti-research extremists lack in truthfulness and intellectual rigor, they make up for in spit and vinegar.

Anti-embryonic stem cell research activists are from the most extreme ends of the political spectrum. They see the issue not in shades of gray, but in stark black and white. In recent years, they have developed a militant bioethics agenda that is radically changing the way American science is funded—by allowing fundamentalist religious extremists to decide what research can be carried out by mainstream scientists at universities like Harvard, Johns Hopkins University, Stanford, and M.I.T. While on the other side of the universe, it would seem, science is embarking on a whole new era of discovery, the American radical right believes that today is its day; now is its moment. From this volatile mix, there is bound to be a clash of historical proportions.

In this book, I hope to clear up some of the widespread misconceptions about this groundbreaking field and to provide readers with an understanding of how the promise of regenerative medicine is being threatened by a new, and highly undemocratic, political order. My goal is to illuminate the basic science—what stem cells are, where they come from, and their true potential to treat disease. But I also hope to provide a better understanding of just how and why this revolutionary science, dropped into today's warring political landscape, came to be hijacked by groups that have used it to serve a larger political agenda. Stem cell research has become today's flash point in the clash between the forces of religious and political conservatism and a brave new medicine, being created in the wake of a scientific revolution. In the latter part of the book, I will explore some of the ethical issues that each person should consider as they decide for themselves whether they support, and would like to be the beneficiaries of, all the forms of stem cell research.

As you consider these issues, I hope you will keep in mind that we are all patients at one time or another. Anyone who lives long enough will almost certainly develop some medical condition that could prove amenable to a cure (or a treatment) based on stem cell research. And if we were to somehow escape the laws of probability and never develop a cell-based disease or injury ourselves, no doubt our loved ones will. Stem cell research should not be viewed as a mere academic abstraction, because it has the potential to move rapidly into the intimate space of the clinic, and to profoundly affect the life-and-death decisions of countless doctors, patients, and families. In recent years, the U.S. government has moved aggressively to make those decisions for us, hindering research and denying hope to millions of patients. I hope this book will illustrate how vitally important it is for all citizens, in any democracy, to understand today's biomedical issues, which stand to profoundly affect each and every one of us. Only through understanding the stakes can we arm ourselves to fight against the tyranny of a powerful anti-science minority.

chapter two

two worlds colliding

If the great story of the last century was the conflict among various political ideologies—communism, fascism and democracy—then the great narrative of this century will be the changes wrought by astonishing scientific breakthroughs.[1]

—*Cynthia Tucker, Atlanta Journal-Constitution*

The religious right around the world has made embryonic stem cell research the surrogate battle between religion and science.

—*Robert Klein, California Institute for Regenerative Medicine*

Anyone following the development of today's cutting-edge medical research could think that they've walked onto a veritable minefield of explosive ethical issues. Aldous Huxley's 1932 novel, *Brave New World*, has been bandied about in recent years at U.S. Senate hearings and at Congressional briefings by those who envision an impending nightmare world with science spinning out of control. Politicians, religious leaders, political action groups, and some of President George W. Bush's closest advisors warn that if scientists have their way, the human race is headed toward, if not total extinction, then toward a world so morally deranged that life will not be worth living.

This rhetoric might sound as though it's uniquely suited to our age because it's tied to the latest in experimental research. But in fact the same arguments have been around for centuries. Tensions between the old, the established and accepted, have been colliding head-on with the new, untried, and unknown since the beginning of recorded history. We forge ahead in our drive to understand everything there is to know about the universe and our place within it, yet conservative forces inevitably resist. Usually, the resistance lies with religious authorities, and opposition is attributed to religious dogma. Sometimes, however, opposition to new scientific developments isn't based on religious scripture. It simply reflects entrenched cultural beliefs and superstitions. In this chapter, I hope to provide a basic overview of the historical clash between medical science and religious and cultural dogma, highlighting a few of the developments that I see as forerunners of today's controversy over stem cell research.

One of the most controversial breakthroughs in the last 30 years was the development of recombinant DNA (more commonly known as genetic engineering), which was born in 1973. That year, researchers Paul Berg and Maxine Singer first discovered how to splice together DNA from a common strain of bacteria and the DNA of a monkey virus, creating an entirely new genetic "entity."[2] The decoding of the entire human genome, completed in 2001, is another breakthrough of historic proportions, and is expected to lay the foundation for an unparalleled understanding of the human body and the role of genes in life span and disease. The other watershed event driving today's biological revolution is the isolation of the human pluripotent stem cell in 1998, a development that opened a whole new universe of knowledge about human developmental biology and how the basic building blocks of life work. In the future, these technologies (and the knowledge gained from them) could be combined to create new treatments and therapies that could radically transform human life and health. Without a doubt, the ethical issues raised by these fields will guarantee job security for bioethicists for decades to come.

The history of medicine is littered with cases that replay certain themes over and over again. The idea that man should not interfere with the supposed will of God has been a major belief throughout history, and is based on one of our most deep-seated religious values. Another major theme is the desire to protect the integrity of the human body, to keep it intact and unviolated by any outside force or influence. This value exerts a powerful influence on us and profoundly affects how we cope with illness, possible disfigurement, and death. This desire to protect bodily integrity has roots in both religious and secular beliefs, and human psychology itself. However, it is also one of the main reasons why there is a chronic shortage of organs to transplant. Some people are still opposed to organ transplantation because they don't understand or accept the finality of brain death. Others secretly fear that doctors will be too quick to let them or their loved ones die if they become organ donors. Others simply can't cope with the idea of cutting into a loved one's body, even if they are dead. If we step back and think things through, most of us would agree that the life-saving benefits of organ donation are a good thing. But that doesn't mean that in an emotional crisis, we would give the go-ahead for doctors to harvest a loved one's organs.

Developments in medical science provide fertile ground for conflict because they tear at the very foundations of what many religions have taught for centuries. The belief that it is God who is in control of human life is fundamental to most religions. For millennia, people have believed that only God can decide when each person will live and die. Indeed, scientists and doctors who delved too deeply into life-and-death matters have often been accused of "playing God," and the history of medicine is in many ways an endless replay of this theme. We have forgotten that many of the routine scientific and medical practices we rely on today originally met with ferocious, sometimes bloody opposition.

In the Middle Ages, the opposition to science came mainly from the Catholic Church. Many of the Church's attitudes toward medicine and healing were ancient even during the Middle Ages, having been taken from classical Greek tradition or from ancient Hebrew scripture. Perhaps one of the oldest ideas in all of human history is the idea that disease and illness are the wages of sin—God's righteous judgment upon those who transgress His laws. Throughout the Old Testament, God is depicted smiting sinners with all kinds of calamities, often including illness and disease.

Humans at every stage of history and in different cultures have grasped at ways to understand disease and death. Even before the formalization of the ideas of sin and punishment into religious dogma, many people universally attributed diseases to supernatural forces—the displeasing of gods or even ghosts. Before the first century, A.D., when the Greeks began to lay the groundwork for scientifically based medicine, the only treatments that medicine men and shamans had at their disposal were the casting of spells and the use of charms and magic. Some illnesses were attributed to possession by demons, and for this disorder, the ancient antidote was exorcism or, even worse, a procedure called trepanning, performed by medicine men. In this procedure, which was prevalent throughout Europe, the Pacific islands, and North and South America, a hole was cut into a person's skull with a saw or a sharp instrument in order to let the demons out.[3] It's unlikely that this cured anything, but on the upside, some individuals actually survived the operation, as evidenced by the discovery of ancient skulls where new bone had grown over the holes. Since there was no science to attribute illnesses to natural causes, the practice of medicine—if it could be called medicine—was in the hands of the priests and other spiritual healers, with their special ability to influence the supernatural world.

In the western world, understanding illness as the wages of sin kept medicine in the hands of priests for century upon century. It was a slow and sometimes agonizing process to bring medicine under the purview of science. The first hospitals of Europe, founded in the Middle Ages, were established by the Church. The main treatments offered by the hospital were blessings and prayers, and ironically, those who were critically ill with a potentially contagious disease were forbidden to enter them. The first glimmers of mathematics and science, developed by the ancient Greeks, were kept alive during the Middle Ages by Arab scholars. These scholars preserved the writings of Hippocrates (c.460–c.377 B.C.), who is considered the father of modern medicine because he began to look to natural causes for disease.

The Church, soon after its establishment as the most powerful institution in Europe, entered into a long and conflicted relationship with medicine. Deeply embedded in the dogma of the Church was the belief that physical suffering, being the judgment of God, should not be circumvented. Because the Church had firmly established itself as the only authority qualified to describe the nature of the universe and to provide guidance in matters of life and death, science was perceived from the very beginning as a threat to its authority. At the same time, for hundreds of years nuns and monks presented the only scholarly class of Europe, and they became increasingly involved with preserving ancient texts from Greece and Rome, which included records of Greek medicine. Later, some even conducted some of the earliest scientific research, and certain sects believed that the healing activities of Jesus impelled them to work toward the amelioration of suffering in the sick.

Among the works preserved by monks were the works of Galen, a Greek doctor who had lived and worked in ancient Rome. Galen had stressed for the medical students of his time that the dissection of a human body was necessary if they were to be properly trained in anatomy. But the Church was bitterly opposed to dissection and vehemently condemned it. Not only did the Vatican consider dissection an act of bodily desecration, Christians (Catholics and later, some Protestant denominations) believed that at the coming of the Judgment Day, the graves would open up and God would literally resurrect the bodies of the long-dead faithful.[4] How could he resurrect those who had been so grievously dismembered?

During the early Renaissance period (usually dated as beginning in the 1300s), Europe began to establish bona fide medical schools, and these schools embarked on the systematic training of doctors. They soon decided that allowing medical students to train by dissecting cadavers made more sense than the alternative—placing live patients at their mercy. So they quietly began to revive the practice of dissection.

In England, there was a steadily growing need for human bodies to train future surgeons, but dissection was so reviled by the public that even in the nineteenth century, there was a near state of hysteria to prevent it. Even when it was performed in the name of scientific progress, many people were convinced that dissection was a grotesque violation of human dignity. Furthermore, they reasoned that if the government took a soft view on the issue, it was but a slippery slope to the point where physicians would let their patients die just to obtain bodies for dissection. Consequently, the law imposed strict regulations on the practice. From 1540 until 1719, British law allowed only four legal dissections to be performed each year, and those had to be performed on convicted criminals who had been executed for their crimes. Dissection was still considered a fate worse than death, and any physicians known to perform human dissections were excommunicated by the Church.

The restrictions on dissections drove the practice underground and opened it up to widespread abuse. It led to a thriving Prohibition-style business in grave robbing and body selling that lasted for centuries. Despite the determination of scientists to continue the dissections, the government of Britain refused to loosen its regulations, and the controversy raged on. This is just one example of what can happen when governments irrationally ban legitimate research and thereby abdicate their role of overseeing it.

Stem cell research in the United States today is in many ways analogous to the treatment of dissections in earlier centuries. Many advocates of the research are deeply concerned that the American government's position on embryonic stem cell research leaves the field open to unethical practices that could damage the ability of scientists to conduct ethical research.

The conflict over dissection was never really resolved until the advent of the twentieth century. By then, there had been a cultural sea change and public opinion had adopted a more scientific worldview. Modern medicine had begun to make great strides, and a general recognition of the good that medicine could do displaced any doubt about the necessity of dissections. The public's horror of violating bodily integrity simply gave way in the face of medical progress that was saving human lives. And the legalization of dissection effectively destroyed the common and illegal trade in bodies.[5]

Believe it or not, inoculation against smallpox and other contagious diseases was also met with intense opposition and cries of "playing God" in both Europe and America. Smallpox was a dreaded disease that covered the faces and bodies of its victims with infected pustules, and it was highly contagious. Periodic epidemics killed thousands of people in one fell swoop, and those who managed to survive the disease were often disfigured for life. In the late 1700s, an English country doctor named Edward Jenner noticed that milkmaids who had contracted cowpox, a less serious infection that could be caught from cows, never seemed to catch smallpox. He reasoned that the cowpox had somehow made them immune to smallpox, and he decided to try deliberately exposing people to cowpox to see if they developed an immunity.[6]

Even though Jenner's solution was highly effective, people were frightened and repulsed by the idea of allowing themselves to be deliberately exposed to a disease that naturally occurred in cows. Those who opposed him put up posters showing people turning into cows, as a frightening preview of what might happen if Jenner's vaccine was accepted. The clergy took an even dimmer view of vaccines like Jenner's. From Europe to America, preachers warned that vaccines were the products of atheists and sorcerers. Smallpox was considered to be God's judgment against sinners, and to try to prevent it was sacrilegious. All inoculations were considered "an encroachment on the prerogatives of Jehovah, whose right it is to wound and smite."[7]

One English preacher railed that Jenner's vaccine was "bidding defiance to Heaven itself—even to the will of God."[8] This controversy continued for more than a century, but over time, Protestants became more accepting of vaccines than Catholics. In 1885, there was a major outbreak of smallpox in Montreal. The Catholics died in droves while Protestants, who had mostly been vaccinated, were barely affected by the epidemic. It was incidents like this that finally forced the Catholic clergy to reexamine their position on smallpox vaccination and on vaccinations in general.[9]

Even the easing of pain was not something that was easily accepted by religious authorities. Another medical advance that met with vigorous opposition in the mid-nineteenth century was the use of chloroform to ease the pains of childbirth. Chloroform had only recently been discovered to be more effective than its predecessor—ether—at rendering surgical patients unconscious, and it had fewer side effects as well. But when the Scottish doctor James Simpson began to promote the use of chloroform to assist women in labor, many churches opposed it. Women were supposed to suffer in childbirth, according to religious officials, who cited Genesis 3:16—"I will greatly multiply thy sorrow and thy conception; in sorrow thou shalt bring forth children." However, by a lucky coincidence, Dr. Simpson was Queen Victoria's personal physician, and when the queen opted for chloroform when giving birth to her ninth child, public opinion swiftly turned and the practice began to be embraced.

The attitudes of the seventeenth, eighteenth, and nineteenth centuries are not so hard to understand when you consider that life then was tenuous and death often capricious. Life expectancies were short and people commonly experienced the tragedies of losing loved ones to a sudden illness or accident. Mothers almost universally suffered the loss of more than one of their children and they themselves frequently died in childbirth, leaving newborn babies behind. A bad crop year could result in widespread starvation, and disease epidemics wiped out the young and the old alike. Scientific knowledge was flimsy at best and offered cold comfort to those who struggled with physical hardships, sometimes staggering grief, and no rational explanation for the human suffering all around them. For most people, their only comfort was in believing that the seemingly heartless "decisions" over who should live and who should die were in the hands of an all-knowing God. And those who fell ill could take comfort in the belief that even if they were suffering the judgment of God, in the spiritual economy of the universe, their "punishment" meant that they would be washed clean of their sins and welcomed into Heaven.

Faith in these tenets had sustained people for millennia, whereas the scientific explanation for things was like a tender sprig of grass just beginning to poke its head up out of the soil. How could one rely on science, when the entire edifice of western civilization rested on the premise that even the most confounding events were the will of an all-wise God? Furthermore, the scientific viewpoint was not very attractive to the average person trying to make sense of his or her life. It was cold and impersonal compared to the view that everything happened in accordance with the divorce will of a loving God.

Time and again the voices of religion and cultural tradition cautioned that only God could make decisions regarding human life and death. Believing that those decisions, however untimely or tragic to the individuals involved, were in the hands of the divine was what enabled many people to cope with what otherwise might have been unbearable. However, as science progressed, it gradually moved much of the control over who lives and who dies into the hands of human beings. Slowly but surely, as scientifically based medicine became more effective at saving and extending lives, the argument against it shifted. Instead of maintaining that it is only God who *has* control over human life and death, the argument evolved into the idea that it is only God who *should have* control over life and death. In other words, mankind may be able to take control of certain events that were formerly out of his hands, but it is sinful to do so. Over time that objection has attached itself to many new medical technologies and practices, and it's still alive and well today.

Rather than abating, collisions between religion and medical science only multiplied in the twentieth century, along with the development of more advanced techniques and technologies. Two developments that caused widespread concern and opposition are within recent memory: organ transplants and the concept of brain death.

Organ transplants, which were pioneered in the 1950s and 1960s, immediately opened up a Pandora's box of controversy. Once again, doctors and scientists were accused of circumventing the will of God by extending the lives of otherwise terminally ill people, but now entirely new arguments came into play. The first heart transplant, performed in 1964, was not the transfer of a human heart into a human being—it was a chimpanzee heart transplanted into a man. The patient lived only two hours, and there was widespread revulsion over the idea of a human being receiving an animal organ. But even human-to-human transplants, which were soon to be performed, excited opposition.[10] Many people argued that the very possibility of organ transplantation cheapened human life to a mere commodity, and would encourage doctors to allow people to die so that their organs could be harvested. Some religions, such as the Jehovah's Witnesses, forbade their followers to be either donors or recipients of organ transplants. (Today the Jehovah's Witnesses allow that the donating or receiving of organs is a matter of individual conscience.)

Organ transplantation catapulted another issue onto the public stage that was a bit hard for many people to understand and accept: the concept of brain death, and the role it was to play in the process of organ donation, retrieval, and transplantation.

The most common sources of organ donation are people who end up in hospitals with massive head traumas incurred in terrible accidents. Someone who has received an irreversible brain injury in a car crash, for example, but whose organs are still functioning thanks to a respirator and other artificial means, would be the most likely candidate for donation— if and when, that is, the person is determined to be brain dead, and the death is accepted as final by the person's loved ones. It's not enough for the patient to have expressed a prior wish, even in writing, to be an organ donor—the family must agree to the idea in order for surgeons to proceed.

While brain-dead patients on life support are the best candidates for organ donation, the decision to inject chemicals into the body that prepare organs for donation and to terminate life support can be agonizing for family members. After all, the person doesn't look dead. With continued blood circulation and oxygenation provided by a breathing tube, the patient appears to be alive, and is warm to the touch. Even if the finality of brain death has been accepted, it is very difficult for family members, who are likely to be in shock over the catastrophic event that placed their loved one on life support in the first place, to take the final steps of turning off the machines and allowing the organs to be removed. Many who come to the decision to terminate life support still cannot cope with the idea of removing the organs, and who can blame them? Even the most pragmatic among us can have trouble letting go of the poignant desire to see our loved one's bodies preserved and cherished to the greatest extent possible. Those who can rise to the need of helping others in the midst of their grief are, without question, acting heroically.

Organ transplants, even though they have been around for 50 years now, still lie in a gray zone between social rejection and acceptance. This has led to a chronic shortage of organs to transplant. If this weren't the case, there would not be almost 90,000 people on transplant waiting lists in the United States alone.[11] Many Catholics and biblical fundamentalists throughout the world are still opposed to transplantation because they believe that Judgment Day will bring the literal resurrection of their bodies. Some other religions believe this as well, and forbid their followers to embrace organ donation. In fact, in Japan, believers in Shinto will not donate their organs because they believe that the integrity of the body is essential after death. As a consequence, organ transplants are almost non-existent in Japan. (During the years of 1997 through 2000, there were only ten deceased organ donors in all of Japan.)[12] And those of us who don't feel restricted for religious reasons might still be unable to overcome the desire for bodily integrity, even after death, in ourselves and our loved ones.

Intensifying the conflicts between modern-day religion and science is the fact that the number of biomedical discoveries is increasing at a dizzying rate. Not only does biological science now have a huge foundation of data to build upon, new technologies are speeding up the process like never before. Computers have made data analysis exponentially faster than it was only a few years ago, and powerful machines like electron telescopes allow scientists to observe life at the molecular level. With the development of nanotechnology, scientists are now learning to manipulate physical matter at the molecular level. Stem cell research is just one of these new break-throughs, and it too finds itself at the center of a maelstrom of controversy in the United States and in several other countries. Stem cell research was born into a family of technologies—assisted reproduction techniques used for infertility—that was already at the apex of some of the most contested and controversial practices of modern times.

Embryonic stem cell research would not exist today if it weren't for the groundwork laid by the science of in-vitro fertilization (IVF). I can still remember the 1978 birth of Louise Joy Brown, the first so-called test-tube baby, born to an infertile couple in Lancashire, England. The news was hailed as a "miracle" in some quarters and decried as the veritable end of civilization in others. Popular magazine covers depicted pictures of test tubes with babies inside, giving the mistaken impression that an entire baby had been grown outside of the womb. There were cries that the new in-vitro fertilization technique would spell the end of the nuclear family because it ripped conception away from the sanctity of the marriage rela-tionship. One of the loudest objections exploited fear of the feminist movement by declaring that there was no longer a need for men, so long as women could obtain sperm from sperm banks and have their own test-tube babies. The Catholic Church condemned the process (and still does) from beginning to end, starting with the way the biological father obtained the semen.[13] There were many cries that science was on a slippery slope to cloning and the creation of "designer babies."

Louise Brown was born on July 25 at Oldham General Hospital in the midst of a full-blown media circus. The fact that Lesley Brown was able to carry her own biological child, and that of her husband, John, was indeed amazing: At the time of her pregnancy, Mrs. Brown had no fallopian tubes. When the Browns first visited Dr. Patrick Steptoe, the pioneering gynecol-ogist who, along with research scientist Robert Edwards, created in-vitro fertilization, 29-year-old Lesley was deeply depressed after almost a decade of trying and failing to conceive.[14]

When she first visited Steptoe, Mrs. Brown had undergone an unsuc-cessful surgery to remove some obstructions in her fallopian tubes that

were preventing her from becoming pregnant. The couple didn't know that
the surgery had been badly botched, and they hoped that with the proper
help, she might still be able to conceive.

Dr. Steptoe was not only one of the originators of IVF, he was also a pioneer of laparoscopy, the method of inserting a lighted tube through a small
incision in the naval to see inside the abdominal cavity. When he performed
this procedure on Lesley, it was evident to him that what was left of her
fallopian tubes was utterly useless. The tubes had been severely damaged by a
previous infection and had a great deal of scar tissue caused by the failed surgery. He promptly removed them. At any other time in the history of
mankind, this would have completely ended any possibility that Mrs. Brown
could become pregnant. Even though she still had functioning ovaries,
human conception—the first meeting of the sperm and the egg—occurs not
in the uterus but in a woman's fallopian tubes. The fertilized egg makes its way
down the tubes into the uterus, where, if all goes well, it will form a connection to the uterine wall. Only then will a pregnancy occur. Mrs. Brown could
still produce eggs, but there was no way for them to travel from her ovaries to
the womb.

By retrieving some of Lesley Brown's eggs and fertilizing them in lab
dishes with her husband's sperm and then transferring them into her
uterus, Steptoe established the first assisted-reproduction pregnancy.
Drs. Steptoe and Edwards had already tried this in a few other women
without success, but with Mrs. Brown, they at last succeeded.[15]

The days following Louise's birth were characterized by cries of moral
outrage over the so-called unnaturalness of the IVF birth. The Vatican soon
issued a statement condemning the process. Many other religions opposed
it as well. Because of the widespread moral objections, Drs. Edwards and
Steptoe had already been turned down for government funding in Britain,
and the U.S. government decided that there would be no federal dollars
applied to the development of IVF techniques in the states. The entire field
was driven into the private sector, and, at the same time, outside of the regulations that would have come with federal funding. Nevertheless, in the
years since, IVF has progressed quietly in private laboratories and at fertility
clinics. Infertility is such a common problem (about one in ten couples
experience it), that there has been no shortage of couples willing to undergo
almost anything in order to have a child of their own. Today, over one million children have been born throughout the world thanks to the technique.

Robert Edwards and Patrick Steptoe could not have imagined, back in the 1970s, that when they managed to create human embryos in lab dishes, this development would open up a whole new frontier, not just in the study of infertility, but in the search for cures to a myriad of diseases and disabilities. Although considerable research had been done on animal embryos, it wasn't until IVF that scientists had human embryos to study. It is because of the relative difficulty of establishing successful pregnancies that thousands of embryos now exist, cryogenically frozen, at fertility clinics all over the world.

Fertility doctors learned early on that unless four, five, or even six embryos were transferred into the womb at one time, a woman's chance of becoming pregnant was pretty low. At least up until very recently, the vast majority of transferred in-vitro embryos simply passed through the patient's body without being implanted. Even when multiple embryos are transferred, a woman undergoing IVF is very lucky if one of them attaches to the womb and establishes a pregnancy. The older the woman is at the time she receives fertility treatments, the harder it is for her to become pregnant. Some women, even after several tries, or "IVF cycles" as they are called, walk away disappointed, without ever having become pregnant. Others face the opposite problem—too many of the embryos implant, and they end up pregnant with twins or triplets.

However, for every couple that has to deal with the implantation of multiples, many more complete their reproductive goals (say, having one or two children) with several embryos left over. In fact, in a formal study conducted in 2002 by the RAND Corporation, a private research organization, and the Society for Assisted Reproductive Technology, it was determined that there were approximately 400,000 frozen embryos sitting in clinic freezers in the United States alone.[16] Because many more couples have entered into the IVF process since then, this number has probably increased substantially, but for the purposes of discussion throughout this book, I will assume that 400,000 is a more or less correct estimate.

As IVF techniques were refined, researchers learned to do some of the "sifting" that nature might do, identifying chromosomal abnormalities and testing for inherited disease mutations. This has given rise to the specialty of preimplantation genetic diagnosis, or PGD. It would be dangerous to assume that every time a sperm cell meets an egg, a perfectly healthy, genetically normal embryo results. Natural conception creates a significant number of embryos that are "genetically challenged" in some way, having either too many or too few copies of specific chromosomes, conditions that can lead to devastating diseases or developmental problems such as Down Syndrome (which results from having an extra copy of chromosome 21). If such abnormalities are so common, one might ask, why don't we see more babies born with such defects? The reason is because in nature, most chromosomally abnormal embryos would never implant in the uterus—they simply pass through the woman's body as if they never existed.

IVF added a new dimension to assisted reproduction when it began to identify which embryos are healthy and most likely to thrive. While PGD does not routinely check for the entire list of human diseases known to have a genetic predisposition (i.e., cancer, Alzheimer's disease, etc.), couples who have their embryos created in lab dishes at fertility clinics can ask that the embryos be tested for certain diseases that run in their families. The technique involves removing a single cell, called a blastomere, from a six- to eight-celled embryo and performing genetic tests on the cell. At such an early stage of existence, the process does not destroy the embryo; if implanted, it could still develop into a normal baby. Today IVF clinics can test for the mutations for a growing number of diseases, including cystic fibrosis and Thalassemia (a hereditary form of chronic anemia). As the gene mutations for more and more diseases are identified, more tests will be available for IVF embryos.

Some people are uneasy with the idea of doing genetic testing on embryos before deciding to transfer them to the womb. They're bothered by the idea that embryos with genetic abnormalities aren't given the same chance as healthy embryos to develop into a baby. It's true that parents are unlikely to transfer embryos that have known genetic problems, but to assume that those embryos would have a better chance in nature is probably not correct.

Others are afraid that parents will use PGD to select for physical attributes such as eye color, height, and athletic ability, but, this is based on a misunderstanding of the complexity of genetics. Those who fear that IVF preimplantation genetic diagnosis is being abused by parents who want to pick and choose their children for such superficial reasons can rest easy, at least for the time being. It simply isn't possible to identify beauty queens or NFL athletes in the petri dish. What *is* possible, and that perhaps should elicit some concern, is that the future *sex* of an embryo can be determined. And given the real-world examples of parents in some societies resorting to aborting female fetuses, this is a legitimate concern, and one that deserves our attention.

So the science of assisted reproduction, which set out simply to help infertile couples have their own children, removed conception from the traditional battlefield of women's bodies and in so doing unleashed an explosion of new issues. Traditionalists have taken issue with the prospect of fertility doctors helping those who would not otherwise bear children circumvent the will of the Almighty. IVF also allows parentage in a number of ways that, strictly speaking, nature never intended.

For example, it allows a woman who cannot produce viable eggs to become pregnant using eggs donated by another woman, making her the child's *birth* mother but not its *genetic* mother. Those donated eggs could be fertilized by a male partner, making him the child's biological father, or it could be fertilized by donated sperm. It allows other women to be the gestational mothers of embryos that are completely unrelated to them and become pregnancy surrogates for women who are unable to carry a child. It makes it possible for couples to screen their embryos for genetic defects, and to decide whether or not to have them. To add fuel to the fire, the practice of deep-freezing embryos and preserving them for many years means that it's also possible that an embryo could be transferred into an unrelated woman and brought to term many years after its creation. In theory, a child could be born 100 years after its biological parents and even its siblings have lived and died.

So what about the 400,000 frozen embryos lying in a state of suspended animation in those freezers? What will be done with them? The existence of these tiny frozen dots has become a contentious political issue, but practically speaking, there are a number of possibilities for their dispensation. Scientists estimate that up to half of them may not survive the thawing process, but even so, that still leaves around 200,000 embryos. Some will be used for reproductive purposes, but there is no doubt that thousands of these embryos are no longer needed for family-building. Their biological parents have already had all the children they want, and it is up to these parents to decide what they would like to happen with their unused embryos.

Most clinics ask potential parents at the outset of the IVF process what they would like to happen to any embryos that remain after they have fulfilled their reproductive goals, and they are offered a list of options. One option is to donate them to other couples who cannot conceive, one is to allow the embryos to be discarded as medical waste, one is to keep them frozen indefinitely, and the other option is to donate them to research. Although no one has kept national records of the number of couples willing to donate their unused embryos for research, it is believed to be much higher than those willing to donate them to other couples. In fact, since 1980 (the year the first IVF baby was born in the United States), less than 100 children have been born from embryos that were donated to couples other than their biological parents.

When embryos are donated for research, most likely the research will be conducted by the IVF clinic that produced them. It is this practice, over the years, that has allowed the field of IVF to make scientific advances. A small number of clinics have relationships with university scientists who do research with them, but a ban on federal funding of embryonic research makes this extremely rare. Remember, IVF has evolved in the United States completely in the private sector because of the lack of federal funding. Because of their dependence on government funding, the vast majority of U.S. scientists cannot work with excess IVF embryos. It's only with private funding that American scientists can use these embryos to do any kind of research on them, and, as will be explored in an upcoming chapter, there is very little private funding to be had.

However, public policy is in a state of flux, and as we look ahead into the future, it's reasonable to assume that at least some of those frozen embryos could end up being used in stem cell research. They just might end up establishing cell lines that are multiplied in the lab, shared far and wide among scientists, used to create specific cell types for the treatment of disease, and one day, end up in you or me.

Clarifying a problem that haunts patients and pro-research groups alike, the first study ever to determine what percentage of fertility clinics dispose of some unused embryos (published in July 2004 by bioethicists Arthur Caplan and Andrea Gurmankin) found that 84 percent of clinics routinely do so.[17] Oddly enough, in today's strange political climate, no one is objecting to those embryos being thrown away, but the fight to keep them from being used to find cures for disease is ferocious. But before we attempt to understand the political cross currents that have led to this ironic situation, it's important to understand the basic science that has spawned such a controversy in the first place.

chapter three

the science that started a revolution

I think that support of this research is a pro-life, pro-family position. This research holds out hope for more than 100 million Americans.

—*Senator Orrin Hatch (R-UT)*

[Allowing embryonic stem cell research] . . . is also likely to lead to human cloning and the harvesting of body parts from babies conceived for this purpose.

—*James Dobson, Focus on the Family*

Science cannot resolve moral conflicts, but it can help to more accurately frame the debates about those conflicts.

—*Heinz Pagels*

One of the most unfortunate byproducts of the stem cell wars has been wide-scale confusion about the science behind the debates. Public opinion is clouded by dramatically opposing viewpoints presented in the media, where scientists and patient-advocates are frequently pitted against conservative political action groups, politicians, and members of the religious right. More often than not, the two sides simply contradict each other to the point that the only result is more confusion.

In addition, there are some people who believe that *embryonic* stem cells and therapeutic cloning are indispensable if we are to usher in a new paradigm in medicine. Yet many of us have been told that *adult* stem cells are sufficient—that they can do everything that's needed to banish a host of diseases. We are told that embryonic research is not only morally unacceptable, it's unnecessary as well.

The soldiers on this side of the conflict also insist that therapeutic cloning, or the cloning of patient-specific embryonic stem cells, is no more than a prelude to the wholesale cloning of human beings. Americans are hearing blatantly contradictory claims about embryonic stem cells versus adult cells almost every day. It's no wonder so many people are confused about what the science actually consists of.

While we're making up our minds about this new science, there's another factor to consider that is just as important to most of us as our health: the matter of a good conscience. In spite of how much we could all be affected by the life-saving breakthroughs of stem cell research, the vast majority of us feel that we should only be heir to research that is morally ethical. Who among us would be callous enough to go through life blithely accepting the idea that his own good health depended on the cruel sacrifice of innocent others? The science of stem cell research is being presented to us in just these terms, but is that really the choice that we have to make? In the last chapter we looked at the clashes between science and religion over the last few centuries, leading up to the creation of embryos in the lab. In this chapter, I'll present some of the scientific facts about stem cell research, before moving on to how this research has been woven into current-day politics.

Stem cell science is unprecedented in the enormous range of different diseases and conditions that it could treat. The reason that it has so many applications is because all of our body's functions depend on the healthy functioning of cells—the tiny engines that drive every process vital to life. As discussed in chapter one these powerful little units perform a staggering array of specialized functions. They convert nutrients to energy, build all the tissues and organs, and make the hormones and chemicals that are necessary for them to function. They form an electrochemical network throughout the body that allows them to communicate with each other, and some specialized cells constantly patrol the blood, tissues, and organs, searching for infections and foreign materials that don't belong there. But like every living thing, cells have a limited life span. They are constantly being born, growing, and dying. While we go about our daily business, cells are quietly tearing down and rebuilding our muscles and bones, sprouting new connections in our brains, and mobilizing here and there to fight off potential toxins and infections. Health is not a static condition but a dynamic equilibrium in which cell birth and cell maintenance outpace cell degeneration and death.

As we learned in the first chapter, cells have a finite life span, and we are only provided with a limited number of them in one lifetime. Young cells divide energetically, with few genetic "copying" mistakes, but as we get older our cells stop replenishing themselves at the vigorous rate of young cells until finally, they can no longer divide at all. When cellular renewal is winding down, we experience the signs and effects of aging, including the development of age-related problems like heart disease and cancer. The principle behind stem cell research is that the formula for self-renewal, and for continuous cellular replenishment, hides within stem cells—the parent cells that generate new cells. If we could unleash that regenerative power, control and direct it, we could cure or reverse a great many catastrophic diseases, injuries, and birth defects that are currently beyond the reach of medicine.

There is overwhelming evidence that the overall curative potential of stem cells is real, and truly revolutionary. Stem cells could bring about an entirely new approach to healing by permanently replacing cells and tissues in the body that have died or stopped working. Living cells could replace drugs as the pharmaceuticals of the future, providing us with our own internal sources of the chemicals and hormones that we need. The healing of damaged organs could be stimulated from within the body by an infusion of healthy stem cells (or cells grown from stem cells) that would multiply into large numbers of replacement cells. By rebuilding diseased organs from the inside out, injections of new cells could eliminate the need for more invasive surgeries and possibly even organ transplants. Stem cells can also provide a way to grow whole new organs that are genetically matched to the patients receiving them. In theory, it might also be possible some day to stimulate the body's production of its own adult stem cells on a scale that would overtake and defeat a number of diseases.

Scientists have discovered several different types of stem cells, and expect they may discover even more. *Embryonic stem cells*, which are produced during a fleeting period of time in the very early-stage embryo, hold the blueprint for every cell, tissue, and organ of the human body. The ability to generate all of those cell types is a quality that scientists call *pluripotency*. In other words, these master cells can generate heart cells, brain, skin, bone, or literally any of the cell types that make up the human body. As time goes by and the cells of the embryo divide, they become increasingly specialized. Farther down the developmental pathway are *multipotent* stem cells, which can give rise to a family of other cell types, but not to *any* cell type.

Multipotent stem cells can also be found in many parts of the fully developed body, including bone marrow, blood, brain, skin, and the gut. At this point they are called *adult stem cells*. Scientists have discovered that our bodies make adult stem cells throughout our lives, and that these cells are intimately involved in healing. Sometimes referred to as "progenitor" cells, while adult stem cells can give rise to other cell types, they cannot convert into a *totally unrelated* cell type.

The claims being made about the medical potential of stem cells sound so dramatic that some people question whether any single field could possibly live up to such soaring expectations. But there is good reason to believe that stem cell research in general holds tremendous medical promise. For example, there is now an established history of patients being cured of leukemias and other deadly blood disorders through bone marrow transplants—the first human (adult) stem cell transplants. While bone marrow transplants have been successful in treating blood cancers (by generating new blood cells) for years, science is only now beginning to catch up to learning how the stem cells in marrow give rise to all the blood and immune cell types, and how this in turn cures disease. And while embryonic stem cell research has a much shorter history, it also rests on a solid, four-decade foundation of animal research, where "proof of principle" has been demonstrated time and again. The ability of embryonic stem cells to give rise to the entire spectrum of cell types has been well established.

Because the opponents of embryonic stem cell research claim that adult stem cells can do everything that's needed to cure disease, considerable confusion has arisen over the relative potential of the two types of stem cells. Those who oppose the use of embryos in research claim, for political reasons, that adult stem cells can be used to obtain any cell type—that adult cells, in short, are pluripotent. However, the possibility that adult stem cells can give rise to even a large number of cell types is far from proven, and a majority of scientists believe that adult stem cells have far less curative potential than embryonic stem cells. While it is *possible* that some day scientists might very well unravel the mysteries of what makes a cell pluripotent, and be able to use that knowledge to return adult cells to that primitive state, we are far from that point today.

To complicate matters, some members of the press have reported several instances in which scientists claim to have converted bone marrow stem cells (the adult stem cells called hematopoietic stem cells) into completely different cell types, including heart cells and neurons. This is often followed by widespread publicizing by right-to-life groups who announce that adult stem cells have been "proven" to be pluripotent. But according to the larger stem cell research scientific community, none of these reports has offered proof that adult cells have been "reprogrammed." So far, not one of the experiments in which researchers initially believed that they had converted an adult stem cell into a completely different cell type has been duplicated, even when it was tried by several different groups. And many of these studies have been completely disproven.

The effort to convert adult stem cells (usually the above-mentioned hematopoietic stem cell) into different cell types has been tried many times by growing the bone marrow cells in cultures where they are mixed with the desired cell type—say, neurons. Scientists hoped that the immature marrow cells will pick up cues from the neighboring neurons, which might stimulate them to differentiate into neurons themselves. One of the most famous examples of this was reported in 2002 by researchers at the University of Minnesota. The adult stem cell researcher, Catherine Verfaille, reported that she had observed evidence that the hematopoietic stem cells she cultured in different cell mixtures were behaving like a variety of totally different cell types, including heart, brain, and liver cells.[1] This would be really exciting news, and it would have an enormous impact on the entire field if it proved to be true. However, when several other research teams, including those from Stanford University, the Howard Hughes Medical Institute, and the University of California at San Francisco, tried to duplicate the results, they were disappointed. What they found was that some of the marrow cells had simply fused with the different cells they were cultured with, creating cells with abnormal numbers of chromosomes. In other words, cells that could have no possible clinical use.

So far, there have been no systematic, peer-reviewed, and published studies showing adult stem cells to be pluripotent. This hasn't stopped pro-life activists, though, from overstating Catherine Verfaille's research. To this day, Dr. Verfaille's early experiments are still being cited as ersatz proof of the ability of adult cells to return to a pluripotent state. Dr. Verfaille herself has complained about this, stating that, "My research is being misused, depending on the point someone wants to get across. They have put words in my mouth."[2]

In the fall of 2004, David Prentice, a senior fellow of the conservative political action group the Family Research Council, testified at a U.S. Senate hearing on stem cell research. The hearing was being held to help U.S. senators understand the potential of stem cell research in light of pending legislation they might be voting on. In it, Prentice claimed that there had been "a wealth of scientific papers published over the last few years" proving that adult stem cells have essentially the same qualities, and the same curative potential, as embryonic stem cells.[3] I met David Prentice in April 2005, when we both spoke at a University of Alabama stem cell research conference. Although the vast majority of scientists have insisted otherwise, he firmly stands by his arguments. He claimed that adult stem cells have cured 56 different diseases, whereas embryonic stem cells have cured none. This is a perfect example of how one can be technically correct in his facts, yet highly disingenuous in regards to the truth.

Prentice left out a key fact. The only cures that have resulted from adult stem cell transplants have belonged within one group of diseases—blood diseases. Leukemias and lymphomas have indeed responded very well to transplants of stem cells culled from the patient's own blood or bone marrow, or from cells found in umbilical cord blood (considered to be an adult stem cell). In addition, patients with autoimmune diseases, such as Crohn's disease and lupus, have responded very well to stem cells taken from their own blood. This experimental treatment is not without its dangers, though: The patient has to have the immune cells in his or her blood completely destroyed through radiation before receiving an infusion of blood stem cells. However, the blood stem cells do seem to repopulate the patients' immune systems with healthy blood and immune cells. The technique of using adult blood stem cells to treat autoimmune disease has yet to be systematically tested on a large scale, independently verified, and published, but so far these results are encouraging.

It looks as though the ability of hematopoietic stem cells to treat many variants of blood and immune diseases is turning out to be a true godsend to those who suffer from these terrible conditions. However, the successes in treating many blood diseases with cells from the patient's own bone marrow or blood, as valuable as they are, do not in any way suggest that adult stem cells are anything more than multipotent. We are still talking about the hematopoietic stem cell, giving rise to blood and immune cells, its own family of cells.

Adult stem cells have also raised hopes for the treatment of damaged hearts. Some encouraging research suggests that stem cells taken from bone marrow, when infused into a damaged heart, can help heal and rebuild damaged heart muscle. This treatment might also stimulate the growth of new blood vessels, creating a kind of "natural bypass" for clogged blood vessels leading to and from the heart. Scientists in Germany, Britain, the United States, and other countries have reported benefits from these trans-plants to patients with heart failure or damage following a heart attack.[4] At first, scientists thought that the hematopoietic stem cells were repopulating the heart muscle with new cardiac cells, but once again, upon closer obser-vation, it was found that the bone marrow cells were not actually differen-tiating into heart cells. However, the stem cell infusion does seem to have a measurable effect on the heart, so there must be some other mechanism at work. Scientists have more recently concluded that, even though adult stem cells can't morph into a totally unrelated cell type, they nevertheless secrete healing factors that help repair the body's existing cells.

Of course, the major advantage to adult stem cells is that they can be taken from the patient himself. There is no possibility of a patient rejecting his own cells, and no need to suppress the immune system (which would put the patient at risk for deadly infections). If adult cells could actually be converted to a pluripotent state, as so many hope, then we would have found the holy grail of stem cell science. Not only would the issue of genetic matching be eliminated, but most of the moral objections to the research would evaporate. Why most, and not all? Because some people theorize that an adult cell, if returned to a pluripotent state, might actually have the remote potential to develop into a complete organism. In adult cells, the genes that play an active role in driving embryonic development have been switched off. To return the cell to an embryonic state would theoretically turn them back on, and then what? What would be the developmental potential of that cell? In certain animal species, single embryonic stem cells that have been implanted into a female animal have resulted in the growth of a complete organism. The possibility for this to happen in humans must be infinitesimal, but it would be naive to assume that right-to-life absolutists would have no objections to the use of pluripotent cells in research.

The bottom line is that adult stem cells have definite limitations. Although they do seem to have healing properties that are not well understood, there are many diseases for which no adult stem cell has been discovered that would generate the needed cell type. But the lack of cell types isn't the only problem. There are technical difficulties in working with adult stem cells that don't present themselves in embryonic stem cell research. Stem cells are very hard to identify in cultures of adult tissue, making it difficult to isolate and grow pure populations of them. (Adult stem cells don't look much different from any other cells in a lab dish.) Also, once isolated, they do not multiply with the same energy as youthful cells, making it extremely difficult to produce enough of them to be therapeutically useful.

An interview with the acclaimed stem cell researcher Irving Weissman was recently highlighted on the Stanford University School of Medicine's website. In it, Dr. Weissman says:

> Several opponents [of embryonic stem cell research] previously have claimed that any adult stem cell could turn into any other tissue, and so neither embryonic stem cell research nor nuclear transfer stem cell research would be necessary. Although this notion has been thoroughly disproven by several independent

groups, the advocates persist in their claims. While we can hope that such disinformation is not accepted by the public, I fear that these claims are now being viewed through the lenses of politics and of the media, and not on the basis of medical or scientific evidence.[5]

So what about the widespread assertion from opponents of embryonic stem cell research that embryonic cells have never cured a single disease? They are correct, of course, because even though scientists have been working with adult stem cells for about 50 years, human embryonic stem cells were only isolated for the first time in 1998. Since then, they've received only a trickle of U.S. government funding, under conditions that are extremely discouraging to scientists who would like to work with them. However, to say that there is no scientific evidence to support the potential of embryonic stem cells is wildly incorrect. It's ironic that those who see such phenomenal curative potential in adult stem cells have labeled the most turbo-charged stem cells useless. There's no question whatsoever that all of the therapeutic qualities that adult multipotent stem cells have, embryonic stem cells have in spades. We know this because mouse embryonic stem cells have been available to scientists since the 1970s, and the work done with them and other animal embryonic stem cells has provided definitive proof that the cells can indeed give rise to any cell type of the body, and that they have a truly remarkable ability to heal and to grow new tissues. Even the limited amount of research done with them in the lab has surpassed expectations.

Embryonic stem cells are derived from five- to seven-day-old embryos called *blastocysts*. They have certain qualities that—at least by today's science—can't be matched. They have a proven ability to give rise to any cell type of the human body. These are normal, bona fide cell types—not fused cells—and they can be produced in very large numbers. Embryonic stem cells proliferate much faster than adult stem cells, and can produce *enormous* numbers of healthy new cells. Cells derived from a *single* embryo could produce enough stem cells to treat a huge number of patients. In fact, Thomas Okarma, who is president of the California stem cell research firm Geron Corporation, has estimated that enough neurons could be derived from a *single embryo* to treat *ten million Parkinson's patients*. In contrast, he says, adult stem cells are seriously limited as a mass-market treatment because too few cells can be grown from a single source to make the treatments effective.[6]

Embryonic stem cells have another advantage: They are in mint condition. Cells taken from the adult body can have damaged DNA that will lay the foundation for cancers and other diseases, but embryonic cells have not suffered any of the wear and tear of life. And they are much easier to identify and isolate from other cells in the lab-created embryos they are derived from. The embryo at the blastocyst stage is a tiny, fluid-filled ball with an outer membrane formed from a thin layer of cells. Inside the membrane, the pluripotent stem cells form a distinct clump, which can be removed and cultured, or grown, in lab dishes. The cells can be nurtured in a highly controlled lab setting, where they can be used to grow large numbers of new cell types for medical treatments. But they have several other uses which are almost as important. They can be used to test new drugs on specific types of cells, greatly reducing the need for testing on lab animals. They can reveal how different agents, including toxic agents, act on human cells without putting human subjects at risk. In one of their less celebrated but hugely important capacities, these cells can also provide a mother lode of information about human developmental biology that can't be obtained in any other way.

But along with the special promise of embryonic stem cells come special concerns that have to be worked out scientifically before treatments become available. One concern is that, because the cells divide so prolifically, will they know when to stop dividing once they've been transplanted into the body? Over-dividing cells could become a recipe for tumors, or the overproduction of chemicals and hormones. Because of this danger, it's very unlikely that stem cells in their *pluripotent* state will ever be transplanted into humans—first and foremost, scientists must learn how to direct them into a desired cell type, whether that cell type is another stem cell of more limited potential (a multipotent stem cell), or a terminally differentiated cell (a body cell with a permanently established identity). Only then can they be safely transplanted into patients.

Another major concern is that cells taken from IVF embryos can be rejected by the body because they are not genetically matched to the patient. The problem of rejection is just as real in cellular transplants as it is in organ transplants, and this is why stem cell scientists believe that therapeutic cloning is so important. So far, *therapeutic cloning* is the only proven way that scientists know of to create embryonic stem cells that are not at risk of being rejected by the patient.

Unfortunately, right-to-life groups have confused the public with as many alarmist predictions about cloning as they have about embryonic stem cell research. Contrary to what many anti-cloning activists want us to think, therapeutic cloning and reproductive cloning are two very different things. *Reproductive* cloning is the creation of an exact genetic copy of an entire organism. This has been achieved in several animal species—Dolly the sheep is an example of a reproductive clone—but to do this in humans is in no way the goal of legitimate stem cell researchers. *Therapeutic cloning*, which is also referred to as *nuclear transfer*, is a technique for creating embryonic stem cells that are genetically matched to a patient. The process of harvesting the embryonic cells destroys the embryo, putting an end to any possibility that a baby could somehow result.

Therapeutic cloning involves taking a human egg cell and removing its nucleus, which contains most of the cell's DNA. The enucleated egg is then fused with a cell, most likely a skin cell taken from the body of the donor-patient (let's call her Sandra). The DNA in the nucleus of Sandra's skin cell becomes the egg's DNA. With the use of certain chemicals and a mild electric shock, the egg is activated to divide. As it does so, each resulting cell will have an exact copy of Sandra's genes. They will be her cells, so to speak, only returned to an embryonic state. Within a few days of cellular division (four to seven days), a tiny clump of pluripotent stem cells will appear inside the egg cell, which is now referred to as an embryo.

If our goal was to create a reproductive clone of Sandra, the embryo would at this point have to be transferred into a womb with the hope that a pregnancy would result. But in therapeutic cloning, this never happens. The embryonic cells are only allowed to divide for a few days, with a strict upper limit of about 200 cells or less. These cells are the pluripotent "master" cells that can give rise to *any* cell type that Sandra needs. Their life span is brief, however. They must be quickly removed from the inner cell mass and put into lab dishes to be cultured. If the cells were allowed to divide for more than about seven days, they would lose their pluripotency and start to differentiate into a mix of different cell types—in other words, they would lose their medical usefulness. So the claims of anti-research activists that scientists are pushing to develop cloned embryos into fetuses and then harvest their body parts are pure nonsense. What scientists are interested in harvesting are these primitive, undifferentiated cells that only exist in the first few days of cellular division.

One of the less publicized but still critically important outcomes of therapeutic cloning is the ability to clone the cells of patients with genetically based diseases. Cloned cells that carry the mutations for muscular dystrophy or cystic fibrosis, for example, can be studied from the embryonic stage on through senescence (or cellular "old age"). This ability to create a "disease in a test tube," so to speak, would allow scientists to study in minute detail how genetic diseases evolve in human cells from their earliest beginnings. But this research, as valuable as it could be, receives no federal funding in the United States and is even in danger of being criminalized if anti-cloning bills (such as one recently promoted by Senator Sam Brownback) become law.

Again, opponents of therapeutic cloning object to the science for two main reasons. One objection is that scientists who claim that they are only seeking to make stem cells are setting the stage for human reproductive cloning. But even if transferred into a uterus, the potential of a cloned human embryo to develop into a baby is open to dispute, and may not exist at all. Scientists attempting to clone other primates have found that critical developmental genes are inevitably turned off in the primate embryos created through nuclear transfer, making it impossible for them to develop past a few days. Another concern is that an embryo created for research could somehow be developed into a fetus. However, this is quite a stretch when you consider that harvesting the pluripotent cells from the inner cell mass of an embryo destroys any possibility that a fetus could develop.

Human embryonic stem cells have been used to generate new blood cells, nerve cells, lung, muscle, cardiac, and immune system T-cells. Again, these are normal cells, not fused cells with genetic abnormalities. Human embryonic stem cells have been proven to reverse paralysis in rats with spinal cord injuries, and this has been independently verified by several research teams, including John Gearhart at Johns Hopkins University and Hans Keirstead at the University of California at Irvine. The effect was not a 100 percent cure, but it was *far* more dramatic than any of the results obtained from using adult stem cells. Both human and animal embryonic stem cells have now been used in animal paralysis experiments, with astonishing results. Rats that could not use their hindquarters at all, after receiving embryonic stem cell transplants, were able to walk on them, albeit with a limp. There have been many remarkable developments in just the past few years that—with sufficient research—could yield dramatic cures for a huge array of diseases.

The following is a chronological list of only a few of the most important developments in research using embryonic stem cells:

November 1998: Human embryonic stem cells were derived for the first time from blastocysts by James Thomson at the University of Wisconsin. The cells were isolated from embryos that had been created at an IVF clinic and later donated to research.[7]

November 1998: At around the same time, John Gearhart at Johns Hopkins University isolated primordial germ cells from fetal tissue. The resulting stem cells are thought to have similar potential to that of embryonic stem cells. These cells have proven to have great experimental value in revealing the properties of stem cells.[8]

December 2001: Two independent research teams, using different techniques, announced that they had produced neural progenitor cells (the parent cells of brain and nerve cells) from human embryonic stem cells. This meant that scientists were on their way to developing cellular cures for a wide range of neurological and brain diseases, including Parkinson's, ALS, cerebral palsy, stroke-related brain damage, and even some day Alzheimer's disease.[9]

January 2002: Embryonic stem cells were injected into mice with Parkinson's disease. The cells turned into functioning, dopamine-producing neurons that reversed the symptoms of Parkinson's. This particular experiment could never have been tried in humans because the cells were pluripotent, but the researchers were amazed that these primitive cells "homed in" on the sites of brain damage and supplied the new cells that were needed.[10]

January 2002: Scientists at Advanced Cell Technology, a private biotechnology firm in Massachusetts, reported that they had produced the first cloned human embryo. The embryo only produced a small number of cells before it stopped dividing, but the work provided further evidence that the cloning of patient-specific human embryonic stem cells is possible.[11]

February 2002: Embryonic stem cells were cultured from monkey parthenotes. A "parthenote" is simply an unfertilized egg that has been activated to divide. Four of the eggs divided up to the blastocyst stage, and the cell lines created from them included neurons and heart cells. This too was accomplished by scientists at the Massachusetts firm Advanced Cell Technology, demonstrating that human parthenotes may be a possible source of embryonic stem cells.[12]

March 2002: Research led by George Daley at Harvard combined therapeutic cloning with gene therapy in treating mice with severe immunodeficiency disease ("bubble-boy disease"). The genetic defect that causes the disease was corrected in cloned embryonic stem cells, which were then differentiated into genetically matched hematopoietic stem cells. These cells were transplanted into the mice, partially rescuing their immune systems. This seminal study showed that therapeutic cloning, combined with gene therapies, has a strong potential to cure genetically based diseases. It means that stem cells may be the long sought-after vehicle for delivering corrected genes into the body, making gene therapies a reality.[13]

December 2002: Scientists transplanted kidney precursor cells from both human and pig sources into mice. The stem cells grew into functioning, appropriately sized kidneys that developed their own blood supply and produced urine. Even though the transplanted cells were not genetically matched to the recipients, these kidneys grown from immature cells were shown to have less rejection risk than fully formed kidneys transplanted from adult mice. This provided intriguing evidence that new organs can be grown for those suffering from end-stage heart, lung, liver, and kidney disease.[14]

June 2003: Human neural stem cells were injected into paralyzed rats and the rats' movement was restored. As I mentioned earlier, in this seminal study, rats that had been unable to move their hindquarters, after treatment, were able to walk again. The team, led by John Gearhart, also concluded that the stem cells, in addition to stimulating the growth of new cells, had "rescued" and restored damaged neurons that might have otherwise died.[15]

July 2003: California scientists reported that they too had injected nerve cells derived from human embryonic stem cells into paralyzed rats, and the treatment enabled the rats to walk again. The transplanted nerve cells, called oligodendrocytes, were found to have formed new myelin sheaths around nerve cells and also to have secreted growth factors that stimulated the birth of new neurons. This study and the previous one have given hope to patients who suffer from paralysis as a result of a spinal cord injury or motor neuron diseases such as ALS that some day, doctors may be able to cure human paralysis.[16]

October 2003: Scientists at the Massachusetts Institute of Technology reported that they had used human embryonic stem cells to grow three-dimensional tissue structures, including liver tissue, cartilage, nerve tissue, and blood vessels. This breakthrough brings us a step closer to being able to grow replacement organs from embryonic stem cells. If scientists can succeed in growing new organs from therapeutically cloned stem cells, the organs would be genetically compatible to patients, radically reducing the dangers of rejection. This feat could also solve the critical organ shortage.[17]

April 2004: Scientists in Israel reported that they had generated pancreatic beta cells (the insulin-producing cells that are destroyed in the bodies of diabetics) from human embryonic stem cells. So far it does not appear that these cells can be derived from adult stem cells, so the ability to generate them from embryonic stem cells is critical. Scientists hope that cells like these can soon be used to replace the insulin-producing cells in the pancreata of human diabetics.[18]

September 2004: A research team led by Israeli scientist Lior Gepstein culled cardiac cells from human embryonic stem cells and transplanted them into the hearts of pigs that had abnormally slow heartbeats. The transplanted cells acted as "cellular pacemakers," regulating the rhythm of 11 of the 13 animals tested. This breakthrough could lead to a natural, cellular replacement for mechanical pacemakers.[19]

January 2005: Japanese scientists reported that they had grown dopamine-producing neurons from monkey embryonic stem cells and transplanted them into the brains of monkeys with a primate version of Parkinson's disease. The new cells integrated with the animal's brain tissue and partially reduced their Parkinson's symptoms. The cells didn't survive well after transplantation, though, and the next step is to determine why and to promote the long-term survival of these cells.[20]

March 2005: Researchers from Ohio State University reported that they had found a way to mass-produce undifferentiated embryonic stem cells by growing them in a bioreactor, a tissue-growing device that is able to exponentially increase the number of cells grown as compared to the number that can be produced using conventional culturing methods. The scientists estimated that they could reduce the cost of multiplying embryonic stem cells by about 80 percent, and the cells grown in the chamber could grow in three dimensions, allowing them to develop into a more natural shape than growth in a petri dish would allow. This brings the technology of embryonic stem cell research a major step closer to treatments that could be developed and distributed in a cost-effective manner.[21]

April 2005: Researchers at the University of Wisconsin-Madison implanted neural stem cells grown from embryonic stem cells into rats with ALS. The scientists reported that the transplants had halted the disease. If this could be carried over into humans, it would be the only known method for stopping the progression of ALS, which today is incurable and inevitably fatal.[22]

June 2005: Researchers at Australia's Monash University reported that they had developed ovary-like structures containing eggs from mouse embryonic stem cells. This research could lead to the development of human eggs for infertile women who produce no eggs, and could also someday lead to an ample source of human eggs to be used in therapeutic cloning without requiring women to become egg donors.[23]

October 2005: Scientists at the University of Minnesota reported that they had coaxed human embryonic stem cells into becoming cancer-killing cells in the laboratory. It has already been established in animal research that stem cells can home in on tumors and cancer cells in the body, so this work helps pave the way for cancer treatments that would act as "smart bombs" in the body, hunting down metastatic cancer cells and destroying them. Because of the "homing" properties of stem cells (adult stem cells also have this ability), they have already been shown to locate hidden cancers in the body that can't be detected by conventional medical tests.[24]

January 2006: James Thomson, the University of Wisconsin researcher who first isolated human embryonic stem cells in 1998, reported that his team had grown two new embryonic stem cell lines in a medium that was completely free of animal cells or growth factors. This is a major breakthrough, meaning that embryonic stem cells can now be produced for human therapies without the risk of transferring animal infections. Thomson believes that he may even be able to decontaminate the government-approved cell lines by "washing" them in his newly developed culture medium, but this needs to be independently verified. Any new cell lines developed using this method will be far more suitable for human transplantation than those derived using the older methods of growing them atop a layer of mouse "feeder" cells and bathing them in animal-derived serums. However, under the restrictions imposed by the Bush administration, government-funded scientists will not be able to work with any newly created cell lines derived using the new technique.[25]

These are only a few examples of what scientists have been able to do already with embryonic stem cells in a field that is itself embryonic. It's hard to imagine what might have been accomplished by scientists in the United States by now if funding for the research were not severely curtailed. The last word (for now) about the need to do embryonic research in addition to adult stem cell research has to do with the hoped-for ability to reprogram adult cells to an embryonic state. If scientists are ever to learn how to recreate the pluripotent state, they need to learn what makes cells pluripotent in the first place, and the only way to do that is through the study of embryos.

When all is said and done, most scientists believe that adult stem cells will be able to cure some conditions, while only embryonic stem cells will hold the cure for others. But even adult stem cells will never reach their full potential until we have unraveled the mysteries of embryonic stem cells, the tiny seeds that hold the intricate master plan for the human body and its innumerable functions from birth to old age.

Of course, we can't assume that stem cell transplants have no risks, and there are many scientific hurdles to be crossed before cellular cures can be delivered to the patients who need them. Some of the hurdles include learning how to direct the differentiation of stem cells into the desired cell types, then ensuring that transplanted cells integrate successfully with the body's own cells. Scientists must learn how to assess whether transplanted cells are working properly in the body, and how to make sure that the cures are long-lasting or permanent. And of course, the issue of genetic matching must be resolved, whether it is through therapeutic cloning or through another approach. These problems can only be solved over time and with adequate funding to keep the field moving forward.

Time and funding—those are the critical issues in the United States today and in several other countries as well. Scientists believe it's not a question of *if* stem cell cures will become a reality, it's a question of *when*. With enough funding, there could be several cell-based cures available in the next five to ten years, but if the United States continues its policy of severe restrictions, many cures could be decades away. Unfortunately, there are millions of seriously ill Americans who don't have the luxury of that much time. In the following chapter, I'll examine some of the political roadblocks that are slowing the development of human cures to a crawl.

chapter four

hijacked by the politics of abortion

I'm in favor of protecting life. And when you take one life to save another life—that's just wrong.

—*Judie Brown, American Life League*

I think it's time that we recognized the Dark Ages are over. Galileo and Copernicus have been proven right. The world is in fact round; the Earth does revolve around the sun. I believe God gave us intellect to differentiate between imprisoning dogma and sound ethical science, which is what we must do here today.

—*Rep. Christopher Shays (R-CT)*

Roe v. Wade will die the death of a thousand cuts.

—*Anonymous pro-life activist*

From the moment the Stem Cell Research Foundation's website went live, phone calls and e-mails poured in from people looking for cures for a bewildering array of diseases, injuries, medical complications, birth defects, and injuries. As the manager of public education, part of my job was to answer questions from the public. However, because of the totally unprecedented number of people contacting the foundation, on many days that was my one and *only* job. Most of them were either sick themselves or were family members trying desperately to get help for a loved one. Many people had reached the end of a long, frustrating journey in which traditional medicine had completely failed them, and after finding the Stem Cell Research Foundation on the web, they looked to it as a beacon of hope. Others sent jarring and hate-filled e-mails accusing the foundation of promoting abortion and the murder of children. In the second part of this chapter I'll explain how this bizarre juxtaposition came to be.

I heard from mothers with brain-damaged children, elderly people going blind from macular degeneration, middle-aged diabetics facing limb amputations, young adults slowly losing control over their bodies to ALS, and husbands desperate to bring their wives back from a stroke-induced coma. It was unsettling just to contemplate the long list of slowly creeping diseases, disabilities resulting from a stroke or a heart attack, and quite a few calamities that stopped me in my tracks, marveling at the perverse blows of fate that could randomly strike anyone at any time. As I ploughed through the desperate e-mails and phone calls, I became acutely aware of how tenuous our grasp on life and health really is.

Two contacts that occurred in 2003 will stay in my mind forever. One was an e-mail from the wife of a man who walked out of the house one cold winter morning, slipped on a patch of ice, and fell onto his back. This is something most of us have done at one time or another, with nothing more serious than a sore backside and some wounded dignity to show for it. But this man—in that one moment—became a quadriplegic. He crushed his spinal cord, and two emergency surgeries did nothing to restore the feeling or movement of his body from the neck down. His wife explained that he had been a very active, energetic person who loved sports and the outdoors. She ended her message by saying that he had expressed a desire to die rather than live in his current condition. In that short, staccato message, I could sense this woman's terror, and her desperation to move forward under the staggering weight of it all. She wondered if there was any hope, any stem cell treatment that could help reverse her husband's paralysis. Sadly, in my reply, I had to tell her that human stem cell treatments for spinal cord injuries are several years away from being a reality in the United States.

Another haunting e-mail came from the mother of a 25-year-old woman who was rushed to the hospital emergency room while in the throes of an asthma attack. The young woman, who I'll call Donna, was struggling so hard to breathe that the emergency room staff forced a breathing tube down her throat to help her. In what must have been one of this mother's worst nightmares, Donna slipped into a coma and was admitted to the intensive care unit, where she was hooked up to a respirator. Then, for some unknown reason, the young woman developed a high fever, which raged on for several days before abating. On her eighth day in the hospital, she opened her eyes, and with the breathing tube still in her throat, she mouthed the words, "Mommy, why am I blind?"

Although Donna's doctors have never been able to say for sure, they think that her over-stimulated immune system attacked and severely damaged her optic nerves. A few years later, Donna still lives in total darkness, unable to even detect light. Her only hope at the moment is that her body will somehow heal on its own, but her doctors have told the family that there could be another answer in the foreseeable future. In a few years' time, Donna's optic nerves might be regenerated with the help of stem cells.

Given this ray of hope, Donna's mother started avidly searching the web to find out more about the revolutionary treatment that might restore her daughter's sight. When she contacted me, she was already aware that the research that might help her daughter had only been performed in animals, but hoped there might be a human clinical trial somewhere to give her daughter hope. Again, I had to tell her that the research just isn't there yet.

Like Donna's mother, most of the patients and loved ones who contacted the foundation were more informed than the average person. They knew there is a promising new field in medicine that could help the condition they wanted to cure. I couldn't give them medical advice, but what I *could* do was pass along information about research that was going on and, if possible, direct people to universities and doctors that might have some answers for them. But there was a painful theme that played itself out time and time again.

For the past few years, scientific evidence of the curative potential of stem cells has accumulated at breathtaking speed. Especially promising are embryonic stem cells. But as shown in chapter three, this explosion of potential still needs to cross some critical scientific hurdles before it is harnessed into safe and effective treatments for humans. And that's a very hard thing to tell anyone who is in a race against time with a disease that is rapidly running its course, or the mother, father, wife, brother, or husband of someone who is dying or in pain.

Patients are now asking, "When will these hurdles be crossed? How long do we need to hang on before we can benefit from cellular transplants?" Unfortunately, the painful message that millions of sick Americans are getting is that stem cell transplants are still in the realm of theory. Research in animals has provided powerful evidence that a cellular transplant could work, but human treatment is not available here in the United States, and no one knows when it will be. The type of research we so desperately need is either not being done in this country or it's proceeding at an excruciating crawl. Here, in the biomedical research capital of the world, the most promising medical research of our time is hindered by the refusal of our government to provide adequate funding for it.

This is not a message that the average American finds easy to digest, or to believe. We take it for granted that the United States leads the world in cutting-edge science and medicine. Most Americans are under the impression that if there is an answer for any disease, disability, or catastrophic injury, as long as we can afford it, we will have that treatment at our fingertips. We may have to do a little searching, dig for information, find the right doctor, treatment, clinic, or even clinical research trial, but sooner or later, we will have access to the best of the best. And for most of the years following World War II, this has been true.

America's lead in the world of biomedical research was not accidental. It has been the result of a more or less consistently pro-science government policy, which has poured billions of dollars into the National Institutes of Health (NIH) over the past half-century. In 2005, the NIH had a budget of $28.8 billion, much of which was funneled to more than 212,000 scientists at universities, medical schools, teaching hospitals, and independent research organizations around the country.[1] For the most part, deciding which research will receive government funding has been left in the hands of scientists—teams of specialists in a given field who review NIH grant applications and decide whether or not they merit public dollars. For decades, this system has kept America at the cutting edge of medicine and technology. However, it is in no way guaranteed that our government will continue to support the most promising research, or that Americans will have access to the most effective cures, or that we will continue to lead the world in science and medicine. In recent years, medicine in the United States has come to depend on the political climate to decide what is acceptable, and which research will receive the generous funding that is necessary in a world of increasingly expensive, technology-driven research. As science continues to extend the frontiers of knowledge into previously undreamed-of territories, medicine is likely to be even more dependent on the shifting winds of politics. Stem cell research is a prime example of how medical science has become a pawn—and many would say a casualty—of modern politics.

Most people have heard of stem cell research—how could they avoid it? The subject has been on the cover of practically every major magazine and newspaper in the country, and is brought up on a regular basis on all of the TV network news shows. Yet misconceptions about how it is proceeding in this country, and even about what the research consists of, run rampant. For the millions of Americans for whom this innovative field represents their best hope of a cure, it is critical to understand how the political environment has all but brought a screeching halt to embryonic stem cell research, the most promising avenue of all.

On the federal level, funding for embryonic stem cell research is mired in a legislative quagmire that has been in a holding pattern since the election of George W. Bush. Conservatives insist that the Bush policy (announced by the president in August 2001) to allow limited funding of embryonic stem cell research is sufficient to move the field forward. But scientists, patient groups, universities, and quite a few politicians from both parties have decried the Bush policy as woefully inadequate. Most stem cell research advocates see the Bush policy as a political sleight of hand that actually does more to inhibit the research than to support it. Both sides of the issue are fighting passionately, with no sign of backing down. Battles over this issue have been vociferous in the press, on the floors of the U.S. Congress and the Senate, and in state legislatures throughout the country. So who is right? Is stem cell research proceeding at a healthy pace, or is it indeed being strangled?

First, let it be said that there is no objection from any quarter to scientists pursuing research on adult stem cells—those that come from the fully developed human body (of any age after birth) or from umbilical cord blood. There is no restriction on government funding for adult stem cell research. Nor is there any restriction on stem cell research in animals. In fact, in 2005, the latest year for which figures for past expenditures are available, the National Institutes of Health invested over $567 million in research using *animal* stem cells and *adult human* stem cells. So those who say that stem cell research is moving ahead in this country are correct when it comes to animal experimentation and adult stem cell research—two areas that are vitally important. However, there are severe restrictions on government funding for research into *human embryonic* stem cells, the cells derived from very early-stage, lab-created embryos called blastocysts. In 2005, the same year that the NIH invested over $607 million in stem cell research overall, research using *human embryonic* stem cells received only $39 million. In other words, the NIH spent about fourteen times as much on animal and adult stem cell research as it did on human embryonic stem cell research.[2] In addition, therapeutic cloning—the creation of an embryonic stem cells using the DNA of a patient, receives *no* government funding whatsoever.

As I mentioned in chapter one, one of the first decisions George W. Bush made following his election to president was to allow a small trickle of federal funding for scientists who want to do research with human embryonic stem cells. But this policy has been deeply misleading, if not outright deceptive, from the beginning. In August 2001, the president announced to the nation that, after a considerable amount of reflection and soul-searching, he would allow federal funding to be used for research on 64 already existing human embryonic stem cell lines. Because of federal restrictions on funding for research that destroys an embryo, these cell lines had been created in the private sector, without the benefit of federal funds. (A cell "line" is simply many batches of cells that were taken from a single embryo and multiplied in lab dishes.) The issue that this decision supposedly hinged on was the fact that human embryos had already been destroyed to create the cell lines, and no public funding was being used to destroy any new ones. In a televised address to the nation, Bush took pains to emphasize that he had labored over his decision, saying, "Embryonic stem cell research offers both great promise and great peril, so I have decided we must proceed with great care. As a result of private research, more than 60 genetically diverse stem cell lines already exist . . . I have concluded that we should allow federal funds to be used for research on these existing stem cell lines, where the life-and-death decision has already been made . . . This allows us to explore the promise and potential of stem cell research without crossing a fundamental moral line by providing tax-payer funding that would sanction or encourage further destruction of human embryos that have at least the potential for life."

The immediate response from many people was to applaud President Bush for what was seen as his almost Solomon-like wisdom in reaching this compromise. Sixty or more cell lines sounded like a lot to work with, and many television journalists assumed that Bush had cleverly satisfied his conservative political base while allowing critical research to move forward at a meaningful pace.

The very next day, however, several major newspapers and the scientific press had a far more sober reaction to the Bush decision. The *Washington Post* called the Bush policy "the most restrictive use of money the administration could have permitted short of a ban." Once the news sank in, the scientific community, including some of the most prominent experts in the field, were taken aback by the claim that there were over 60 human embryonic stem cell lines available for U.S. scientists to work with. Of the cell lines that were listed on the NIH website, the vast major-ity were not proven to be pluripotent, the critical feature that makes embryonic cell lines useful. As noted earlier, many of them had not been accurately characterized at all, and some existed in foreign countries that forbid their exportation (making it impossible for cell batches to be sent to scientists at American universities). Still others turned out to be owned and patented by private research companies that didn't agree to share them. Within days of the president's speech, the number of available cell lines on the NIH website began to fluctuate. At first the number increased to 78, as if to buttress the president's case, but as scientists and journalists made inquiries at the labs allegedly holding them, the number of lines rapidly diminished. For one reason or another, the number of embryonic stem cell lines that U.S. scientists actually had access to was chipped away at until reality sank in—of the original 64 cell lines that were claimed, there would never be more than 22 or 23 available, with government funding, for American scientists to work with. And each cell line comes from a single embryo, meaning that all of the cells produced from it will be genetically identical. So in addition to all the aforementioned limitations, the number of approved cell lines falls far short of the range of genetic diversity that scientists say are needed for research and treatments.

Scientists choosing to work with these cell lines are also hamstrung by byzantine funding restrictions, patents that severely limit the uses of the lines, and exorbitant costs in obtaining their use ($100,000 per batch of a single cell line for private companies). In addition to that, allowing government-funded scientists to work only with cell lines created before August 9, 2001 is highly undesirable from a scientific standpoint. It has meant that the technology used to isolate and grow the cells, was frozen at a very early stage of development, and any improvements in the derivation or culturing of embryonic stem cells would be of no benefit to U.S. government-funded scientists. This immediately became a problem when scientists noted that all of the approved cell lines had been grown atop a layer of mouse "feeder" cells. The mouse cells were needed to supply the stem cells with growth factors that would keep them dividing in lab dishes, but that also meant that the stem cells had been subjected to possible rodent diseases. And it meant that the cells had probably absorbed molecules from the mouse cells that the human body would have a severe rejection reaction to if they were ever implanted into a human. In other words, the cells were contaminated and would never be suitable for human transplantation.

The absurdity of the cutoff date has only become more apparent over time, as scientific advances are continuously occurring in the private sector and overseas. In March 2004, Douglas Melton, a researcher at Harvard, announced that his team had created 17 new human embryonic stem cell lines (using only private money), which he was prepared to share with other researchers for free.[3] Perhaps even more importantly, researchers in Singapore have since announced that they have grown human embryonic stem cells without using mouse feeder cells or any animal growth factors whatsoever. Since then, others have succeeded in doing the same.

The elimination of mouse feeder cells and animal growth serums from the technique of growing human embryonic stem cells was a major breakthrough, one that further highlighted the inadequacy of the Bush policy. By the spring of 2005, scientists were reporting that over 120 new cell lines had been created using newer, better techniques that did not involve animal cells or growth serum. However, these cell lines, which have the potential of some day being used in human transplants, remain off-limits to U.S. government-funded scientists.

In May 2005, the U.S. Congress approved a bill that would at least modestly expand the federal funding of embryonic stem cell research. Supported by most Democratic Congressmen and many Republicans as well, the Stem Cell Research Enhancement Act (HR 810) makes no fundamental changes to the president's current policy. What it does do is eliminate the arbitrary date of August 9, 2001 as the cutoff date for the creation of embryonic stem cell lines that can be studied by NIH-funded scientists. It is intended to allow federal funding to be applied to 120 newer, superior cell lines that have been derived in the private sector using excess IVF embryos. However, the bill would impose another cutoff date, the date of its enactment. If it is passed, no new cell lines created after the date that the bill is signed into law will be funded. But the Bush administration isn't about to let even this modest proposal go forward without a fight.

On the day that Congress passed HR 810, President Bush held an elaborately staged press conference featuring a roomful of adorable babies and toddlers with their parents. All of the children were "adopted" as frozen embryos. "There's no such thing as a spare embryo," the president declared, beaming as he hugged a tiny, red-haired toddler. When reporters asked about his reaction to the passage of HR 810 that very day, he said, "I will veto it." He urged Americans to "allow" all the frozen embryos sitting in IVF clinic freezers—now estimated to be about 400,000—to be adopted, rather than be used for research. He seemed either unaware or unwilling to acknowledge that it is the genetic "parents" of the embryos (the egg and sperm donors), not the U.S. government, or anyone else for that matter, who have the power to decide what can be done with them. He had no suggestions for where he expected to find adoptive parents for all 400,000 embryos, even though the number of couples willing to donate their left-over embryos to other couples (and the number of people willing to "adopt" them) is quite small.

If the White House's reaction seemed a little theatrical, the rhetoric from opponents of HR 810 on the floor of the House of Representatives was scalding. Democrats and moderate Republicans called for passage of the bill by appealing to its promise for alleviating the pain and suffering of millions of living people, while some conservatives likened the passage of HR 810 to opening the very floodgates of Hell. When Tom DeLay, then the Republican House Majority Leader from Texas, addressed the Congress, he delivered an incendiary mix of science fiction fantasy and Old Testament—style fury. He referred to the passage of HR 810 as a "moral catastrophe" and "the first drop of the deluge" that would usher in a nightmarish world that embraces the wholesale murder of babies and "a black market in human body parts." This was quite a statement, considering that when it comes to the frozen blastocysts, there are no human bodies, never mind identifiable body parts.

Conservatives in the U.S. Senate immediately mobilized to try to fend off ratification of the bill by their colleagues. In order to siphon off support for the Senate version of HR 810, Sam Brownback (R-KS), Rick Santorum (R-PA), and others soon had at least six competing bills on the table, some of which are completely unfounded in science and are based on pure ideologically based speculation (this will be discussed in greater depth in chapter six).

One of the cruel ironies of the political dance is that only a small minority of Americans wants to put a complete ban on embryonic stem cell research. In the past few years, numerous public opinion polls have been conducted by university researchers and independent research groups. The percentage of Americans who are in favor of loosening the funding restrictions (in other words, increasing funding) for the research has remained fairly consistent, regardless of how the questions are asked or who asks them.

According to an Opinion Research Corporation International Poll, released in March 2003, 67 percent of Americans supported embryonic stem cell research, including therapeutic cloning research, and wanted the government to allow it to go forward. A little more than a year later, support had climbed even higher due to the death of former President Ronald

Reagan and the high profile stem cell research was getting in the 2004 presidential race. Nancy Reagan's courageous call for more funding for the research that might have assisted her husband in this struggle against Alzheimer's struck a cord with a sympathetic public. A survey commissioned by the Civil Society Institute, published in June 2004, found that 74 percent of Americans were in favor of allowing more flexibility in the funding of the research.[4]

The idea that the issue breaks cleanly down political party lines, or even down liberal/conservative lines, has not been borne out by in-depth opinion polls. While liberals, in general, tend to have the most supportive view of the research, polls that break the population down by political party or by religious affiliation have had surprising results. An April 2004 poll published by Peter D. Hart Research Associates, which was conducted in presidential "political battleground" states, found voters who support expanded federal funding to outnumber those against it by 65 percent to 17 percent. Even more interesting about this poll is the fact that, of the churches that have the most extreme official positions on embryonic stem cell research, large percentages of their members do not agree with them. Seventy percent of the Catholics polled were in favor of the research, and among Protestant evangelicals, 46 percent favored expanded funding, as opposed to only 30 percent who were against it.[5]

Nancy Reagan's public support of expanded research using embryonic stem cells not only focused a spotlight on the issue after her husband's death, it had a measurable effect on public attitudes. This was clearly evident in the above-mentioned Civil Society Institute poll. The poll not only reflected the high level of trust the public has in Nancy Reagan, it confirmed what research advocates had been saying all along—that the more people understand about embryonic stem cell research, the more they are in favor of it. The survey found that not only were 74 percent of Americans in favor of the research, 62 percent of self-identified conservatives were in favor of it.[6]

What's more, the stem cell research issue continues to shift farther and farther into bipartisan political territory. The Republican Main Street Partnership, a nonprofit centrist organization, released the results of a poll in May 2005 that showed that Republicans support expanded federal support for the research by a margin of 55 percent to 38 percent.[7] The following month, the *Wall Street Journal* published the results of its own poll, done in conjunction with Harris Interactive Health-Care Polls, showing that 60 percent of Republicans and 83 percent of Democrats favored more funding for embryonic stem cell research. And an increasing number of prominent Republican politicians have broken with the Bush administration's position on this issue, including Senators Orrin Hatch, Arlen Specter, John McCain, and even Senate Majority Leader Bill Frist.

No matter how you slice it, a firm majority of Americans want to see all types of stem cell research go forward with more backing from their government. How is it possible that year after year, the clear will of the majority is ignored, while a small minority (one-third or less) is able to dictate policy to the rest of the nation? This situation is a testament to the power of the right-to-life lobby.

The right-to-life movement has been around for well over a century and has emerged as a powerful force in politics. In the 1800s, anti-birth control and anti-abortion groups focused on the control of women and their choices, not on the embryo or the fetus. However, in the twentieth century, women gained new rights, and central to them was the right to control their reproductive destiny. Once society overwhelmingly accepted the idea that a woman should be able to exercise choices when it came to her own body, the focus of the anti-abortion movement changed. Attention was shifted more and more toward defining the fetus, then the embryo, and now, in extreme cases, even a dividing egg, as a person. Some of the groups who are fighting today to ban embryo experimentation are the same groups, or their modern-day incarnations, that fought against the legalization of birth control in the nineteenth and early twentieth centuries. They have just repackaged themselves as pro-family, anti-abortion, and, to sell their message to a society that is less and less receptive to a blatantly anti-feminist message, "pro-life."

Judie Brown of the American Life League (ALL) is a perfect example of how right-to-life activists have learned to adapt their message in a changing social climate. Considered by many to be the "grandmother of the modern pro-life movement," Brown appeared in a 2004 two-part television news segment about stem cell research that was broadcast in 38 states by the Sinclair Broadcast Group. If you listened to Brown's words, you could easily be alarmed. "The Americans really just want to clone humans. They want you to believe that they're interested in curing diseases, but in reality, curing disease is at the very bottom of their priorities," she says matter-of-factly (even though she herself is American—go figure!). Brown founded the American Life League in the 1970s and is a frequent guest on local and regional radio and TV talk shows, where her latest crusade is against stem cell research and therapeutic cloning. She seems to have her finger on the pulse of a vast left-wing conspiracy this time led by depraved scientists—to stamp out every value that civilization holds dear, especially the reverence for human life.

Although the ALL is best known for fighting the legalization of abortion, Ms. Brown has never really left her anti-birth control roots. She has spent the last few decades protesting not only abortion, but all forms of birth control, and has worked to ban sex education in the schools. Over the years, however, the ALL has stayed true to its origins while adapting to the ever-changing social landscape that now includes smaller families and millions of working mothers. In 1998, the ALL widened its net in response to many of the most recent scientific breakthroughs and gave birth to a more contemporary-sounding think tank called the American Bioethics Advisory Commission (ABAC). The ABAC states that its concerns include "stem cell research, cloning, reproductive technologies, euthanasia, genetics, eugenics and personhood."[8] But this does not mean that the ALL has abandoned its anti-abortion platform—far from it. Through ABAC, the American Life League's anti-abortion agenda has simply been reinvented under the more modern rubric of bioethics, by advocating for the recognition of the embryo as a full-fledged human being. The ALL's blend of tenaciousness and adaptability is a perfect example of how a group that was considered beyond the political fringe 30 or 40 years has become a part of the national political establishment today.

Collectively, conservative social activists have fought hard to get for-right-wing politicians elected throughout the country, and they now have an influence on U.S. government policy that is far out of proportion to their numbers. At a meeting of the Coalition for the Advancement of Medical Research in July 2005, I was present at an address made to that group by Congresswoman Diana DeGette. DeGette is one of the sponsors of HR 810. She highlighted the fact that the network of right-to-life lobbyists has been extremely effective at influencing members of Congress. "There are Congressmen who, when the National Right to Life Committee calls and says, 'vote this way on this bill,' they do it," she said. So how did stem cell research become so entangled in the politics of abortion in the first place?

The idea that embryonic stem cell research is somehow predicated on abortion, or that scientists need aborted fetuses to conduct it is a misconception that has been vigorously promoted by those who want to ban abortion. There really is only one connection to abortion, but it is certainly not one of cause-and-effect. Some universities and other research organizations bank fetal tissues (along with all kinds of human tissues) that have been preserved after either miscarriages or induced abortions, and scientists who are affiliated with those organizations can obtain permission to use them in research. The decision to donate the tissue to research is made by the mother. Scientists who wish to do fetal tissue research are not involved in any decision about whether or not to terminate a pregnancy—they enter the picture long after the woman has made that decision. Researchers who work with these tissues can obtain from them fetal germ cells—stem cells that are derived from what would eventually become the sex cells (future eggs and sperm) of four- to six-week-old fetuses. These cells have been found to be just as versatile (or almost as versatile) as embryonic stem cells, and there are a few researchers in the United States who are conducting research in creating various cell types from them.

Make no mistake about it—these are *not* the cells that are referred to as embryonic stem cells. Embryonic stem cells come only from very early-stage embryos, within the first few days of cellular division and long before a fetus could develop. And it should also be clear that fetal tissue research is legal in the United States and is not actively opposed by the groups that are seeking to ban embryonic stem cell research. Even the Catholic Church poses no opposition to research on fetal tissues. However, the embryos that are used in stem cell research could not have been aborted because they have been created in the lab and have never seen the inside of a womb. Embryonic stem cell research could go forward, full speed ahead, without there ever being another abortion. You will never hear this admitted by any of the pro-life activists, because they hope we will believe that embryonic stem cells come only from abortions, and that the research somehow promotes (or even requires) abortions.

The American Life League is joined by a powerful network of political and religious organizations that have worked relentlessly to draw a connection between stem cell research and abortion. Organizations such as the

National Right to Life Committee, the U.S. Council of Catholic Bishops, the Family Research Council, Concerned Women for America, the National Association for the Advancement of Preborn Children, and Do No Harm believe that if they can ban all embryo research, it will guarantee a ban on abortion. It was almost inevitable that these groups would get involved in the stem cell research issue: Human embryonic stem cells were first isolated in the lab in 1998, when the American right-to-life movement was not only organized, it was gassed up and running on all cylinders.

The issue of human embryonic stem cell research came along at a critical time in the history of the pro-life movement. In 1973, the U.S. Supreme Court ruled, through its decision on the landmark case *Roe v. Wade*, that pregnancy in the first trimester is a private matter between a woman and her body. The decision guaranteed that no state law could prevent a woman from obtaining a legal abortion in the first 12 weeks of pregnancy if she chose to have one. Over the next 35 years, however, activists found that *Roe v. Wade* meant that although states couldn't ban abortions outright, it did not mean that they couldn't enact laws to restrict access to them. Furthermore, the decision protects a woman's right to an abortion only in the first trimester. In the second trimester (after 12 weeks), states can heavily restrict abortions, and in the third trimester, they have the right to ban it. After 1973, anti-abortion groups quickly began to introduce laws at the state level (and to lobby state lawmakers to pass them) that made it more difficult for women to get abortions.

Since the passage of *Roe v. Wade*, there have been several important cases before the U.S. Supreme Court that have defined the limits of the law. The Court has upheld the states' ability to impose many restrictions, but frustratingly for anti-abortion activists, it has always stopped short of overturning *Roe v. Wade*. Time and again, the Supreme Court, with Justice Sandra Day O'Connor considered the critical swing vote, refused to strike down *Roe v. Wade*.

Abortion is not the only issue that many right-wing groups have concerned themselves with, but in the 1970s and beyond, it became a flash point that united a number of political action groups with conservative Catholics and evangelicals, who adopted abortion as their signature issue. In the last 35 years, dozens of conservative political action groups have sprung up that honed their strategies specifically working to ban abortion. One thing they have been very good at is raising money; in 2005, the Women's Bioethics Project, a nonpartisan public policy think tank, estimated that American right-wing lobbying organizations have approximately $300 million in their coffers for advancing their agendas.[9] And those agendas, for the past few years, have become more and more focused on stopping embryonic stem cell research. The big budgets of these organizations mean continuous media outreach activities and media tours for their spokespersons, and the ability to flood the phone lines, e-mails, and fax machines of politicians, telling them how they would like them to vote on legislation. It has also helped them hire hundreds of attorneys to bring lawsuits against legislation they oppose and to engage high-powered lobbyists to work to influence lawmakers in Washington and in the capitols of every state.

These strategies have been quite effective, and right-to-life organizations have scored several victories in making it more difficult for women to obtain abortions. Most state legislatures have been agreeable to passing restrictions on first trimester abortions. And they've been so successful at closing family planning clinics that today 87 percent of counties in the United States have no abortion provider.

With all their incremental victories, the pro-life movement still hasn't been able to achieve a blanket ban on abortion. However, there is one possibility that, if they could pull it off, could bring about a final victory: *To have an embryo from the moment of conception legally defined as a full-fledged human being*. If an embryo were legally defined as a person, abortion at any stage would be considered murder. *Roe v. Wade* would go out the window, and abortion could be criminalized from coast to coast. This is where things stood in 1998, when James Thomson at the University of Wisconsin-Madison and John Gearhart at Johns Hopkins University both (separately) isolated the first human embryonic stem cells. The search for a legal precedent conferring full personhood on the embryo was already well underway, and the research immediately became a magnet for a galvanized movement.

Today even the tiny amount of embryonic stem cell research funded by the National Institutes of Health is under assault. The focus of pro-life activists over the past few years has been to have a state—any state—pass a law that explicitly rules that an embryo is a full-fledged person. An example of such an effort has come from the National Association for the

Advancement of Preborn Children (NAAPC) in Hagerstown, Maryland. This group has tried repeatedly to sue the U.S. Department of Health and Human Services (HHS) on behalf of a symbolic frozen embryo they have dubbed "Mary Doe." In 2004, "Mary Doe" was listed as the plaintiff in a case that sued the HHS for recognition of her "equal humanity and personhood," stating that all embryo experimentation "results in the certain and sudden death of Mary Doe," and "violates her Fourteenth Amendment rights to equal protection and due process of law." The court was asked to issue an injunction against the HHS and the National Institutes of Health, "ordering them to cease and desist any and all plans to undertake human embryo (stem cell) experimentation."

In 2004, the State of Maryland dismissed the suit, but in 2005, in response to the passage of California's Proposition 71—the ballot initiative in which voters approved state funding for stem cell research—the NAAPC filed a similar suit in California. This lawsuit (along with the lawsuits of the California Family Bioethics Council and other anti-abortion organizations) has succeeded in delaying the state's implementation of Prop 71. This time "Mary Doe" even acquired a middle name and became "Mary Scott Doe," and she not only declares that the planned stem cell research threatens her life, she claims that the proposition would subject her and "other human embryos to involuntary servitude and slavery." In both lawsuits, Mary Doe refers to herself as being "born" at the moment of her test-tube conception. As extreme as this suit is in its claims, it has helped to hold up the issuance of the state bonds that are needed to raise money for stem cell research, while the matter plays out in court. For the time being, at least, a petition filed by a "person" who doesn't even exist, except in potential, seems to have circumvented the will of the majority of voting citizens in California. And her "needs" are receiving, at the moment, a higher priority than those of living, actual patients suffering from devastating diseases and injuries.

As the controversy over the fate of the frozen embryos intensifies, several observers of the Bush policy have started to point out one glaring inconsistency in the president's position. How can anyone, they ask, maintain that full-fledged human life begins at conception and at the same time allow current IVF practices to go on? If a blastocyst is truly equivalent to a fully developed person, then IVF doctors and the "parents" of embryos are guilty of murder when they bring about the destruction of embryos, whether it's by freezing them, discarding them, or by using them in research. IVF nurses, technicians, and other clinic workers would be complicit in murder, and the transferring of multiple embryos into a uterus knowing that most of them will fail to implant is also called into question. At the very least, just having thousands of embryos sitting in frozen limbo or being routinely discarded is an embarrassment to the absolutists who are determined to prevent them from being used in biomedical research. Unless thousands upon thousands of surrogate mothers step forward and allow all of the frozen embryos to be transferred into their wombs, the president's stated position can be nothing more than political posturing.

Aubrey Noelle Stimola, who is assistant director of public health at the American Council on Science and Health, wrote in May 2005 that, "Given the number of children worldwide currently awaiting adoption, this seems an unlikely and impractical scenario. . . . If [Bush] truly believes that the destruction of embryos is wrong, shouldn't he attempt to shut down fertility clinics, or at least insist on the immediate perfection of IVF procedures to prevent the accumulation of more unused embryos?"[10] Ms. Stimola also touches on a subject that almost no one has brought to bear on the ongoing debate about "saving" the frozen embryos—the subject of millions of children worldwide who are already born and who are in need of adoption. In the United States alone, over 126,000 children are now in the foster care system, waiting to be adopted.[11] Yet no one has held a press conference showing the world how valuable, adorable, and deserving of our protection they are. The extreme inconsistency of political priorities in this matter brings to mind a statement coined by Congressman Barney Frank some years ago to encapsulate the attitudes of some extreme right-wing politicians. The slogan found its way onto a bumper sticker which I once saw while driving in Atlanta. It said, satirically, "Life begins at conception, and ends at birth."[12]

The pro-life activists know, of course, that any law equating a fertilized egg with a person is bound to be vigorously challenged in court. Their hope is to appeal the case all the way to the Supreme Court and have the highest court in the land uphold it. A decision conferring legal personhood on an in-vitro embryo would compel a revisiting of *Roe v. Wade* and would pave the way to the nationwide outlawing of abortion. With the recent changes in the Supreme Court that resulted in the addition of John Roberts and

Samuel Alito, both conservative-leaning Bush administration appointees, many believe that ambition may not be so farfetched.

But we have to ask if even the most ardent pro-lifers are prepared to deal with the wide-ranging implications of giving full human rights to an early-stage embryo. If this happens, it won't just impact a woman's right to obtain an abortion, it would seriously impact accepted practices in in-vitro fertilization. For starters, such a legal designation would take the "owner-ship" of IVF-created embryos from the egg- and sperm-donating couple and transfer it into the hands of the government. IVF couples would no longer have the right to donate their leftover embryos to research or to have them discarded (any more than they could "donate" one of their children to research or wash them down a drain!). Their only choices would be to attempt to implant all of their embryos, keep them frozen indefinitely, or allow other couples to use them for reproductive purposes—in other words, to allow other people to raise their genetic offspring (assuming that other couples could be found who are willing to do so).

The threat to infertility treatments is no mere speculation. Peter Samuelson, who is president of the nonprofit law firm Americans United for Life, is quite clear on the point that IVF clinics are being targeted by his organization. In November 2005, he told PBS's *Frontline*, "Right now we're working on in-vitro fertilization, an area that's completely unregu-lated, and we've realized there are issues there. . . . There need to be limits on the number of embryos that are created and implanted. And so we're engaged in the conversation with different allies, with lawyers, with legis-lators, and we're drafting a piece of model legislation. We'll take that out next year [2006] and give it to our friends and allies . . . and hopefully within a couple of years, we'll have that passed in some form or in several different states."[13] If he's right, this will be unwelcome news to the 10 per-cent of couples who struggle with infertility, and who look to IVF as their only hope of having their own genetic child.

Legally defining the embryo as a person could have far-reaching effects that even its most passionate supporters have probably not thought of. It would make *all* embryonic research, including embryonic stem cell research, illegal (not just unfunded) throughout the country. At best, it would allow research done with the already existing embryonic stem cell lines but would make it impossible to use excess IVF embryos in research. It would also go far beyond ending abortion and would threaten certain forms of birth control—the RU 486 "morning after" pill, of course, but also most intrauterine devices (IUDs), which prevent implantation of an embryo in the uterus, but not conception itself. If the most extreme interpretation of the law prevailed, even the birth control pill may not be left unchallenged. This is because some birth control pills are thought to prevent not fertilization, but embryo implantation in the lining of the womb. It's not clear how these collateral issues would be dealt with if the incremental successes of the pro-life movement, which have had a cumulatively profound impact, continue unchecked.

One thing is certain, and that is that the powerful pro-life political machine, having proven its mettle in the fight against abortion, is incredibly well tooled to fight embryonic stem cell research at every level of government. The tactics that have won battles in state legislatures are now being applied to fighting stem cell research. The only difference is that so far, the anti-research movement has not employed the traditional conservative slogan of defending the rights of states to enact laws that challenge those of the federal government. At the moment, the anti-research movement has the federal government on its side, and is seeking to *prevent* the states from funding the research in defiance of the federal government. But the battle-tested tactics of the right-to-life movement can be used either for or against the federal government, and the movement has been more than happy to abandon the issue of states' rights when it is expedient to do so.

Patients and their advocates take heart that in 2008, a new president, through an executive order, could reverse some of the restrictions placed on stem cell research by the Bush administration. But given the well-established strategies of the pro-life movement of persistently attacking their issues from every angle and at every level of government, it is naive to think that a new administration will end the war over stem cell research. Several major obstacles now block embryonic stem cell research from proceeding at a meaningful pace in the United States. Federal funding restrictions imposed by President Bush are only one issue. Another legislative issue is restrictions imposed by the Dickey Amendment, which has been part of the Department of Health and Human Services appropriations bill since 1996.[14] This amendment, which was passed by Congress, prohibits the use of federal funding for research using embryos. Other obstacles include a Draconian bill called the Brownback Bill, introduced by Senator Sam Brownback of Kansas.[15]

The Brownback Bill, if it succeeds, would be the most punitive anti-research legislation ever passed in the United States. It would bring one of the most promising avenues of stem cell research—therapeutic cloning—to a screeching halt. The Brownback Bill (titled the "Human Cloning Prohibition Act") makes the cloning of patient-matched embryonic master cells for research and treatments a felony. It would subject scientists and doctors using this technology to up to ten years in prison and fines of $1,000,000. But it doesn't stop there. Patients receiving therapies in which therapeutic cloning was used at any step along the way in developing their treatment would also be felons subjected to the same prison times and fines as the doctors and nurses who helped administer their treatments. And if it is to be strictly adhered to, American citizens who go overseas to obtain life-saving cellular transplants in which therapeutic cloning was used would be subject to arrest as soon as they put foot back on American soil—with the same possible prison time and fines. This bill, which has already passed in the U.S. Congress, is awaiting a vote for ratification by the Senate, which has been deadlocked over the issue for the past three years. But even if the Senate refuses to ratify the Brownback Bill, "mini-Brownback bills" are being introduced in state legislatures across the country, and have even been passed in several states. In fact, in Wisconsin, the home state of stem cell research pioneer James Thomson, a law making therapeutic cloning a criminal felony passed in the state, only to be vetoed in 2006 by Governor Jim Doyle.

Meanwhile, research is slowed to a snail's pace, and the private investors who might provide an alternative to federal funding are shunning embryonic stem cell research, knowing that at any time, the whole field could be shut down, and their investment could go up in smoke. To sum it up, in July 2004, Dr. Gerald Fischbach, a prominent stem cell researcher and dean of the Faculty of Medicine at Columbia University Medical Center, told the *New York Times*, "When you begin arguments based on convictions that are not open to scientific discourse, the whole process starts to crumble, and that worries me, not only with stem cells but with the whole sphere of scientific enquiry . . . there are more and more regulations of science for political reasons. I think it is very threatening. I think it is as threatening as any time in my lifetime, including the McCarthy era."[16]

chapter five

the battle for hearts and minds

The world is a dangerous place, not because of those who do evil, but because of those who look on and do nothing.

—*Albert Einstein*

The only way of finding the limits of the possible is by going beyond them into the impossible.

—*Arthur C. Clarke*

Bernard "Bernie" Siegel had no idea of the wheels he was setting into motion on New Year's Eve, 2002, when he drove to the Broward County courthouse outside of Miami. The tall, silver-haired attorney with the boyish face could look back on 30 years as a successful trial lawyer. With a lovely wife and two children, one in college and one about to begin college, life was good. Bernie describes himself, with self-deprecating humor, as "an excellent lunch companion," because of the incredible breadth of his anecdotes, which he tells in such an engaging way that practically everyone he meets becomes an instant comrade. But Bernie's real gift lies in being able to see the really big issues of our world in a way that few people can. His ability to reach out to total strangers, some of them giants in the fields of academia, government, and public policy, has led him to the epicenter of one of the most inflammatory and contentious issues of our time—human cloning. I started working for the Genetics Policy Institute (GPI), which Bernie founded in 2003, on a full-time basis in 2005, after we had already collaborated on several projects. I had followed his career trajectory for the past couple of years, and in doing so, I was struck by how profound a difference one very determined person can make.

In some ways, Bernie's transition from a private attorney to the founder and head of GPI, a nonprofit stem cell research advocacy organization, was a natural progression. In the 1980s, he served on the board of a nonprofit group called Children's Rights of America. He had learned the ins and outs of family law and became involved with the plights of families with missing children. Since then, he had balanced an interest in civic activism with a busy law practice. Life was proceeding along quietly until December 2002, when a child suddenly came to his attention who, it seemed, was being callously used to serve the purposes of a strange religious organization called the Raëlians.

Only days before Bernie's trip to the courthouse, the Raëlians grabbed headlines throughout the world with the startling announcement that they had been the first in the world to successfully clone a human baby. The news media scrambled to cover the story, and a woman named Brigitte Boisselier, who identified herself as the scientific director of Clonaid, the Raëlians' research lab, appeared on CNN and all the network news shows. The soft-spoken Boisselier was striking to look at. With her soulful brown eyes and dark, flowing hair (which gave her more than a passing resemblance to Morticia Addams), she told the world in a charming French accent that, thanks to the efforts of Clonaid, the first cloned child (named Eve) had been born to an infertile couple.

If it was true, the news would be of historic proportions. From the beginning, the research community had its doubts about the Raëlians' claim because of the sheer difficulty scientists had encountered in cloning all animals, especially mammals. In addition to that, reproductive cloning is an enormously inefficient and expensive process. It takes hundreds of tries to produce one cloned animal that is a complete genetic copy of another animal. Most clones die before birth, and the few that make it to birth are riddled with severe illnesses caused by genetic "omissions" and mistakes. Critical genes, sometimes many of them, are not copied from the donor animal's DNA, and their absence causes catastrophic health problems. By 2002, these disturbing cases of very sick cloned animals were becoming known to the international scientific community and had convinced even Ian Wilmut, the Scottish scientist who cloned the first mammal—Dolly the sheep—that the technique should not even be tried on humans. Still, Dolly existed, as did "CC," the first cloned cat; some cloned cows; and many cloned mice and pigs. The procedure "worked" in a limited way, so no one could say with certainty that Clonaid hadn't in fact produced a very sick, extremely handicapped, but living human child.

After the initial surprise of the announcement started to wear off, some journalists began to voice the concerns of both the public and the scientific community in their interviews with Boisselier. Where was the baby? they asked her. Who were the parents or, in this case, the parent? What was the baby's condition? Over the next few days, Dr. Boisselier assured the world that DNA tests would soon be conducted to prove that the child was indeed a clone. But no matter how much they hammered her with questions, she would not reveal a single detail about the case. As far as the troubling questions about the ethics of performing what was, in essence, dangerous experimentation on a human child, Boisselier insisted that Clonaid had simply provided an invaluable service to an infertile couple. "It's a baby," she said, dismissing their concerns. "It's a cause for celebration."

In Miami, Siegel was following the Clonaid story with great interest. He had learned about the terrible dangers of reproductive cloning a couple of years earlier while helping his daughter on a high school term paper about the cloning of Dolly. He was struck by the fact that those who allegedly produced the child, including her parents, seemed to have a complete lack of concern for her welfare. In his mind, anyone reckless or cruel enough to clone a human baby was by definition not fit to care for her. So Bernie did what no other human being on the planet had thought to do. He drove to his local courthouse and filed a motion asking the court to evaluate the child's welfare, and to determine whether her rights and interests were being protected.

A short time later, as Bernie and his wife, Sheryl, were curling up to watch the news on CNN, they could barely believe it when the network started reporting a major legal development in the case of the "first cloned baby." It took a few minutes before it sank in that reporters were talking about Bernie and the action he had just filed. Apparently, the ink was barely dry on the legal forms when a clerk at the court had picked up the phone and called CNN.

His phone rang off the hook as TV network journalists called to get his commentary. They sent limousines to whisk him to the local stations so that he could appear by satellite feed on the national news shows. Within 24 hours, he had appeared opposite Brigitte Boisselier on CNN and had even debated Raël, the leader of the Raëlians himself, on *Connie Chung Tonight*. The Canadian Raël appeared on camera in full UFO-cult regalia, dressed all in white with his hair in a Samurai-style topknot. Both he and Dr. Boisselier seemed happy to be getting all the attention, but they stubbornly refused to disclose any more than they had said all along— simply that a cloned baby had been born and that they were responsible for her birth.

It didn't take long for the Raëlians' story to start unraveling. Over the next two days, they complained bitterly that Siegel, for some incomprehensible reason, was trying to tear the newborn baby away from her mother. Under searing public scrutiny and a tidal wave of criticism from the scientific community, they used the legal action as an excuse to withdraw their offer of a genetic test to prove the child's origins. Meanwhile, the press was having a heyday with exposés of the Raëlians, showing that their members believe their leader is from another planet and that he came to earth partly to bring immortality to humanity through cloning. By the end of the week, the Raëlians had become fodder for late-night comedians and catnip for angry letters-to-the-editor writers.

As the dust over the Raëlian drama began to settle, Siegel became more concerned that the group was casting a dark cloud of suspicion over legitimate medical research. Inadvertently, their publicity stunt had brought another issue to the forefront—the issue of therapeutic cloning. Having a flamboyant group like the Raëlians making claims of having cloned babies only added to the confusion surrounding the research. It gave the opponents of stem cell research great ammunition to make their case that therapeutic cloning would lead inevitably to the cloning of babies. Bernie decided to persist in challenging the Raëlians, so that he could publicly lay their claims to rest. Before all was said and done, he would also disprove forever the axiom that one person cannot make a difference in a world grinding away under the momentum of large, powerful institutions.

From the filing of the original motion, the judge decided that the case had enough merit to hold an arraignment hearing, which was scheduled for January 24. The normal purpose of an arraignment is to bring all the interested parties, including the parents of the child, together for an initial assessment. At that point, if he thinks it is necessary, the judge could take the child into protective custody while her case was being investigated. But since no one outside of Clonaid knew where the baby was or who her parents were, the judge subpoenaed Brigitte Boisselier and the company's vice president, Thomas Kaenzig, to appear at the hearing and supply more information.

The developing drama was intensified by the decision of Thomas Kaenzig to come to Florida to seek financial investors for his company at a January 11 investment conference in Fort Lauderdale. Siegel had caught wind of the fact that Kaenzig was going to give a keynote speech, and he showed up at the conference with court papers in hand, including a witness subpoena for the upcoming arraignment. The visibly shocked Kaenzig was handed the papers just before his scheduled speech and it was apparent that the last thing he had expected was a real court case in which he would be sworn to tell the truth.

On the day of the arraignment, reporters from around the world crowded into the courtroom to report on the "case of the cloned baby." The room was packed with TV cameras, and CNN even broke into its programming with live coverage of the hearing. Clonaid had engaged two top criminal defense attorneys to represent them, and they showed up for the hearing, without Kaenzig, Boisselier, or any other representatives of Clonaid. Bernie had to resort to questioning the disembodied voice of Thomas Kaenzig, who had arranged to appear by telephone from Las Vegas. For over an hour, he tried to penetrate the shroud of secrecy that had cloaked Clonaid from the very beginning, but Kaenzig, in spite of his lofty title, seemed to know almost nothing about his own company. He admitted that Clonaid was not registered as a business anywhere in the world.

He also said that he didn't know where Boisselier, Clonaid's president and scientific director, was at the time, but he thought that she might be in Canada. The Clonaid lab was looking more and more to Bernie Siegel like a sham organization whose mission was to con desperate couples and gullible investors out of their money. Clonaid didn't seem to have an address or a board of directors, but it did have a legal strategy—to prove that the Florida court had no jurisdiction over the case, and to have it dropped.

Judge John Frusciante wasn't happy with the company's evasions about how it operated and its possible role in a dependency case. What's more, the company was circumventing the court's ability to proceed any further with an assessment of the child's welfare. Without knowing where she was, the judge couldn't even determine whether his court had jurisdiction in her case. He ordered another hearing, and this time he ordered that a Clonaid officer with firsthand knowledge of the case show up, or the company would face charges of contempt of court.

The following week, Clonaid's attorneys filed one motion after another to try to have the case dismissed, without success. A second hearing took place on January 29. Again Kaenzig stayed away, but this time Brigitte Boisselier showed up flanked by Clonaid's attorneys. When she walked into the courtroom, all heads turned her way, and she and Bernie locked eyes. Her manner was smug and defiant. She even seemed to be enjoying the attention, but Siegel was thinking, "Aha, now I've got you, and I'm going to expose you."

With Boisselier at last on the stand, Bernie asked her the location of the baby.

"I refuse to answer," was her reply. He repeated the question several times, and each time she said, "I refuse to answer."

Finally, Judge Frusciante lost his patience and ordered her to answer the question.

"She's in Israel," was all Boisselier would say.

Bernie asked her where Clonaid's lab was and where the supposed cloning took place, and this time the Clonaid attorneys objected. They made their case that the Florida court had no jurisdiction over the child because she was not born in Florida, nor had she ever been to the state. In the end, the judge had to dismiss the case, but he did so with reluctance. "I congratulate you on your resourcefulness," he told Bernie, "but there's nothing more this court can do. If you want to go to Israel, you can pursue the case there."

This was the end of the Florida legal chapter, but it was the beginning of a *much* greater journey. Once Clonaid refused to allow genetic tests to prove that the child was a clone, their credibility was more or less destroyed. The organization still exists and claims to have relocated its "laboratory" to outside the United States. This was done after the U.S. Food and Drug Administration, while investigating the company's claims, found that their original laboratory was no more than a rented room in an abandoned West Virginia schoolhouse. Since relocating their lab to a secret location, they have made claims of cloning even more children, a feat that no reputable scientist has ever come close to. The organization remains as shadowy and elusive as ever. I once asked Bernie Siegel if he was ever actually in the same room with the group's leader, Raël. He said, "No, and it's a good thing, you know. His followers think an invisible UFO hovers around him all the time, and it might have zapped me."

As outrageous as the Raëlian case was, it was only a prelude to the real story. Its lasting influence was to pique Siegel's interest in the issue of stem cell research and therapeutic cloning. The more he learned about the disease-curing potential of therapeutic cloning, and the huge array of diseases it could treat, the more a fascinating, multifaceted story unfolded before his eyes. He looked up the work of the world's foremost scientific authorities, including Ian Wilmut at the Roslin Institute in Edinburgh, Scotland and Rudolf Jaenisch at the Massachusetts Institute of Technology. The scientists he encountered universally thanked him for exposing the Raëlian scam, and soon he learned just how damaging such a fiasco could be to legitimate research. As he delved deeper into the issue, he also learned about the massive campaign of right-to-life groups to spread misinformation about therapeutic cloning.

A large array of these groups were actively working to confuse the issue with reproductive cloning, and the conservative politicians they supported were pushing, and in many cases, passing, blunt-instrument anti-cloning laws that would strike down legitimate medical research. What's more, in the United States, there was essentially no national legislation to regulate either kind of cloning—reproductive cloning or cellular research cloning, never mind laws to protect the rights of cloned people (should they exist). It was a wide-open legal frontier, where almost anything that happened would be a first-ever precedent with the potential to change the course of history.

In fact, the deeper one digs into the cloning issue, the more complicated it becomes. Animal reproductive cloning is being undertaken in many countries, and there is no international law in place to ban that practice from carrying over into human experimentation, or even to set guidelines. In 2003, however, there were two competing international treaties concerning cloning that were being floated at the United Nations. A treaty that had been introduced by Costa Rica sought to ban all cloning worldwide. This treaty, which was being heavily promoted by the Bush administration and by American pro-life organizations, made no distinction between the cloning of *babies* and the therapeutic cloning of *cells*. The Costa Rican treaty was supported by about 50 Catholic and, mostly, developing nations, many of which depended on economic aid from the United States.

However, there was a competing treaty that had been introduced by Belgium and was supported by Britain, France, Germany, Japan, Russia, China, and most of the industrialized world. The Belgian treaty, as it was named, called for a worldwide ban on reproductive cloning, but sought to allow each individual nation to decide for itself whether to allow therapeutic cloning for the development of stem cell treatments.

These two treaties were on the table at the U.N. for two years without ever coming to a vote, and for a while it looked as though the issue was at an impasse. However, in 2003, proponents of the Costa Rican treaty were lobbying members of the General Assembly—those who would cast their country's votes—and lobbying hard. Siegel described the situation to me later as one of unilateral pressures on the voters. "Right-to-life groups were bombarding delegates with pictures of aborted fetuses and the like," he said, regardless of the fact that therapeutic cloning has nothing to do with abortion. They were doing everything possible to try to equate the blastocysts created by therapeutic cloning with babies being aborted. To make matters worse, President Bush threw all the influence of the American presidency behind the Costa Rican treaty. In several speeches made in 2002 and 2003, he had already called for a ban on "all forms of cloning." In his 2002 State of the Union address, he told the nation, "As we seek to improve human life, we must always preserve human dignity. And therefore, we must prevent human cloning before it starts."[1] He made no distinction between the cloning of an entire person and the cloning of cells. He also made his position known to the delegates of the U.N. by having administration officials formally circulate a statement in which the United States called for the ratification of the Costa Rican treaty.[2]

The Vatican also circulated its position papers to the U.N. and enlisted its own delegate, Archbishop Celestino Migliore, to deliver a speech to the General Assembly in which therapeutic cloning was likened to murder. The archbishop went so far as to say that, "The early human embryo, not yet implanted into a womb, is nonetheless a human individual, with a human life, and evolving as an *autonomous* organism [the emphasis is mine] toward its full development. Destroying the embryo results in a deliberate suppression of an innocent human life."[3] I'll return to the issues raised by this statement in a later chapter, because so much of it is open to argument, but when I first learned about this, I could hardly believe that the archbishop used the word "autonomous" to describe a pre-implantation cloned embryo. One can only assume that the inclusion of that word was an oversight, and not a belief on his part that the microscopic clump of undifferentiated cells that makes up the cloned blastocyst could have any life independent of being implanted in a womb. At any rate, the Vatican made its point to the U.N. General Assembly.

Another pro-life group, the Family Research Council, presented David Prentice, the senior fellow of their organization whom I mentioned earlier, to the U.N. Legal Committee as a member of the Costa Rican delegation. It's not clear how this American citizen suddenly became a part of a foreign delegation, but having him presented as such enabled him to address the influential committee with a speech calling for a ban on all cloning, including research cloning. The Family Research Council describes Prentice as "an internationally recognized expert on stem cell research and cloning." However, what he is truly recognized for is not his scientific expertise, but his ideological opposition to all embryonic stem cell research, including therapeutic cloning. He is one of the right wing's most valuable anti-science scientists, and he was allowed to go before the U.N. Legal Committee and argue for the passage of the Costa Rican treaty.

As Bernie Siegel followed the U.N. story, he was struck by how unjustly one-sided the lobbying was concerning the passage of cloning legislation. With the anti-research groups so well organized and vocal, and with the deep pockets of the Catholic Church and groups like the Family Research Council, there was a chorus of voices passionately defending the "rights" of the cloned embryo. Where were the voices defending the rights of living, sentient people, who were suffering from diabetes, heart disease, Parkinson's, cancer, and MS, for whom the research represented their best hope of a cure? Bernie thought that the millions of patients—perhaps over a billion worldwide—who had the greatest stake in the outcome of the U.N. action, the scientists who were in the best position to judge the nature of the research and the mainstream religious community needed to have a voice at the U.N. as well. But how? Here you were talking about a "constituency" that was fragmented, unorganized, untapped. Not only did they need to be *united* in the effort, someone needed to open doors at the U.N. and provide a way for them to be heard.

Siegel soon realized that, the way things were going, a worldwide ban on some of the world's most promising biomedical research was going to be decided on the basis of an emotional campaign waged by groups that were only using the issue as a stand-in for abortion. This was a problem of literally global proportions, but as Siegel explained, "Lawyers are paid to solve problems. So I started asking myself, what would it take to solve the problem?" In addition to linking therapeutic cloning to abortion, the opponents of stem cell research were also disingenuously linking it to reproductive cloning as a way to create a panic among the delegates to ban all cloning. Soon it became apparent that a very helpful first step to getting some reasonable regulations in place would be to separate the two types of cloning. One way to do this, and to slow down the rush to shut down promising research before it had even begun, was to enact a law that banned only reproductive cloning. Therapeutic cloning could then be handled under a separate regulation, clearly delineated and considered for what it is—vital medical research.

By then, though, Siegel realized what a powerful anti-research coalition he was up against. It would not be easy to divert the broad rush of events that all seemed to be leading toward the ratification of the Costa Rican treaty. He told his wife that he wanted to take 90 days off from his law practice in order to focus on the problem. Then he immersed himself in international law and the inner workings of the U.N.

After completing a considerable amount of research, he came up with a simple plan. He would file a petition with the U.N.'s International Court of Justice, otherwise known as the World Court, asking them to issue an advisory opinion that reproductive cloning violates human rights and is therefore a violation of international law. It would be only the second time that a nongovernmental organization petitioned for an advisory opinion from the World Court. The first time was in 1994, when the World Court was asked to provide an opinion on whether the use of, or the threat of the use of, nuclear weapons is ever legal under international law. It took two years and the coordination of a colossal international network of individuals and organizations to bring about the decision. But in 1996, from its headquarters in The Hague, the World Court issued the opinion that there is no legal justification for the threat or the use of nuclear weapons, and that indeed all nations have a legal obligation to enter into negotiations that will lead to their elimination.[4]

Even though the World Court does not have the means to enforce such a ruling, the weighing in of this world body, which represents a majority of the world's nations, carries enormous moral weight. It influences the policies of many nations, and a ruling by the court forever remains as the international legal standard. The nuclear arms decision was a stunning example of how a modern grassroots initiative could actually overrule the world's most powerful governments and have the rights of common people recognized by the highest court on earth.

When Bernie reviewed this case, it became clear to him that if a coalition of nongovernmental organizations could push for, and obtain, a statement on nuclear weapons, why couldn't it be done with cloning? A ruling that reproductive cloning is a violation of international law would seriously damage the ability of research opponents to confuse it with therapeutic cloning. He drafted a petition, and then he started reaching out to the various constituents who would have an interest in cloning legislation. This included patient groups and healthcare nonprofits, bioethicists and legal scholars, and scientists in the field. He wanted to know from the most knowledgeable scientists where they stood on the proposed cloning legislation, so he decided to look up the biggest names in animal cloning and in stem cell research.

He spoke with Rudolf Jaenisch, one of America's most prominent stem cell researchers, then Ian Wilmut, asking them if they would support a ban on human reproductive cloning. They heartily agreed that it should be categorically banned. Even if there were no objections to the idea of violating human uniqueness, they said, the medical problems demonstrated in all cloned animals were currently insurmountable. At least with any of the techniques that are known today, the attempted cloning of a human baby would be a crime against humanity. On the other hand, they also felt strongly that therapeutic cloning to obtain patient-matched embryonic stem cells was far too promising to be ignored. To thoughtlessly ban *it* at the same time would be a tragedy. Both scientists gladly endorsed Siegel's efforts at the U.N. What's more, they helped him make contact with the world's top scientific authorities on stem cell research and cloning. He got in touch with all of them, and there wasn't a single dissenting opinion: Therapeutic cloning should remain legal while human reproductive cloning should be banned.

Soon Siegel had some of the world's foremost stem cell scientists on the advisory board of a nascent organization he was calling the Human Cloning Policy Institute. John Gearhart at Johns Hopkins, Larry Goldstein at the University of California in San Diego, and Douglas Melton at Harvard were among them. Next, he began recruiting legal and bioethical scholars to a human rights legal advisory board. Louis Guenin at Harvard, bioethicist Laurie Zoloth at Northwestern University, and the retired World Court judge Christopher Weeramantry were all enlisted to help advise and guide the U.N. effort.

Finally, Bernie sent the petition he had drafted to Kofi Annan, the secretary general of the U.N., asking to have it adopted by the General Assembly and sent to the World Court for consideration. He circulated copies of the petition among the various missions (the officials from various countries who serve in advisory roles). Many of the scientists he had spoken with wrote letters of support, and Bernie went to work connecting the dots between a myriad of patient-advocate groups who could also endorse the effort. He contacted groups like the Parkinson's Action Network, the Christopher Reeve Paralysis Foundation (now the Christopher Reeve Foundation), and newborn groups like the Stem Cell Action Network (SCAN). The number of independent organizations representing people with the broad array of disorders that could be amenable to genetically matched cellular transplants was in the hundreds. The small foundations advocating for research for children's neurological disorders, kidney disease, or multiple sclerosis, for example, had little power working in isolation. But if they could be united into a single, unanimous voice, each one could multiply its influence exponentially.

Through reaching out to the patient-advocate groups, Siegel met countless people who were struggling to live with crippling conditions. Some of them knew that they would die before research provided them with an answer, but they chose to spend the time they had left fighting for the cures that would save others. Through meeting people like Idelle Datlof of SCAN, who is confined to a wheelchair because of multiple sclerosis, Steve Meyer, an extraordinarily committed Tennessee activist who is in a race against early-onset diabetes, and Sabrina Cohen, a young woman from Miami who has been a quadriplegic since the age of 14 because of a car crash, faces were put to the conditions that the anti-research activists so pointedly ignore. Bernie, a cancer survivor himself, became increasingly angry about the deceptive hype that was holding up the research that might cure some of the terrible diseases that so inexorably ruin and prematurely end human lives.

By then it was the summer of 2003. Siegel's 90-day hiatus from his law practice had long expired, but he was more passionately driven than he had ever been in his life. "I was at a crossroads," he says. "I thought, I can return to my law practice, or I can do something to benefit mankind. I was a driven person, convinced that I could do something to make a difference." This time he decided to devote the next year to fighting any U.N. actions that would have a chilling effect on stem cell research worldwide. Sheryl Siegel saw the importance of what he was doing and supported him despite the fact that his decision would have a major financial impact on their family. By then she was accompanying Bernie on what had become a daily roller coaster ride, with a future that was now a complete unknown.

As the fall approached, the Costa Rican delegation started pushing for an October vote on their cloning treaty. The push for a vote was vigorously backed by the Bush administration, which beefed up its stealth campaign to elicit the votes it wanted. Bernie learned from his contacts at the U.N. that the administration was even threatening to withhold U.S. aid to some less developed countries if they failed to vote for the Costa Rican treaty. It was clear that the Bush administration's leverage over the poor countries could easily add up to a victory, and Bernie knew he had to do something quickly to mobilize the other side.

The few weeks in the run-up to the October vote on the cloning treaty was one of the most frantic periods of Bernie Siegel's life. Every day, he got up at 5:00 A.M. and worked until late into the evening to notify everyone he could think of that it was time to take action. He spent countless hours on the phone calling the U.N. missions, speaking to diplomats, and asking them to reject the Costa Rican treaty and allow the individual nations to decide for themselves on the issue of therapeutic cloning. He e-mailed every constituent group he could think of and had his scientific board, the patient groups, and activists send letters and faxes to the U.N., urging them to keep therapeutic cloning legal.

A major problem in Siegel's eyes was the fact that the international media seemed unaware of the upcoming vote and the profound impact it could have on the average person. The small amount of media attention that had been given to the matter presented the case as being in a hopeless stalemate. Reporters seemed unaware that the Bush administration was hell-bent on passing a treaty that would ban therapeutic cloning world-wide, regardless of the effect this would have on millions of desperately sick people.

Siegel issued a press release alerting the media to the impending vote, and the chilling effect that the Costa Rican treaty would have on legitimate biomedical research. He asked the patient groups to weigh in by issuing their own press releases, and many of them did so. The patient-advocates from around the country immediately jumped in to help. They under-stood how critical it was for the delegates to hear their side of the story, but without Bernie's lead on how and when to contact the U.N., millions of American patients would have had no voice concerning the vote.

He contacted the Coalition for the Advancement of Medical Research (CAMR), the pro-research umbrella organization of universities, medical associations, and patient groups in Washington, and asked for their help. These groups are not generally focused on the proceedings of the U.N. But CAMR's then-president, Michael Mangianello, recognized how the far-reaching effects of the Costa Rican treaty could influence the votes of Washington politicians who would be dealing with similar bills in the U.S. Congress and the Senate. He urged the approximately 90 member organizations of CAMR to call, fax, and e-mail the U.N., stating their organization's position on the importance of banning reproductive cloning but keeping vital research legal.

At the same time that the pro-research community was making its voice heard, pro-life groups stepped up their lobbying efforts with the General Assembly, equating therapeutic cloning with abortion and murder. Insiders at the U.N. were telling Siegel that, although the strong-arm tactics of the Bush administration and the pro-life groups had initially predicted an easy win for the Costa Rican treaty, the delegates were now paying attention to the pleas of the patient groups and health organizations. Only a few days before the vote, it appeared that the outcome was going to be very, very close. Finally, with multiple patient organizations issuing press releases alerting the world about the upcoming vote, the press took notice. Two days before the October vote, both Reuters and the Associated Press released stories about it on the news wire, and this got the attention of newspaper reporters in the United States and other countries.

On the day of the vote, Siegel was sitting in front of his computer in Florida, in a state of almost intolerable suspense when calls started coming in from U.N. delegates who had taken him into their confidence. The General Assembly had decided that the issue of cloning, which had seemed like an open and shut case, deserved more careful consideration than it had received. They decided to put the vote off for the next two years. Clearly, the activist community had raised enough doubts about the merits of the Costa Rican treaty to prevent its passage, but the decision was carried by a razor-sharp margin: The decision to postpone was won by a *single vote*.[5]

Bernie later told me that staving off what had seemed like the inevitable passage of the chilling international treaty was the crowning achievement of his life. "That was the real beginning of the stem cell research grassroots community," he said. "On that day, the tail wagged the dog." But there were no illusions that the stem cell wars were over. If anything, it was obvious that the fight to allow embryonic stem cell research to go forward was going to be a long-term struggle. "The foes of the research have taken this up as a proxy for the abortion debate worldwide, but the research that's at stake represents a paradigm-shift that will impact our lives like nothing else. The U.N. battle has just been the first battle for the hearts, minds and souls of the 21st century."

As many pro-stem cell groups feared, with continued pressure from the Bush administration, the cloning issue came up again in the fall of 2004. Those who had been lobbying to pass the Costa Rican treaty didn't want to lose momentum in their fight for a worldwide ban by waiting another year for a vote. But this time, the research advocates saw the assault coming. The stem cell research activist community, due in large part to Bernie's efforts, had been united behind a common cause.

Earlier, in June 2004, Bernie's growing organization, which had by now been renamed the Genetics Policy Institute, partnered with the Stem Cell Action Network (an affiliation of patients and other volunteers) to host a meeting of nationwide stem cell research activists in California. It was the first time that grassroots activists (most of them patients or the loved ones of patients) from throughout the country gathered together to learn from scientists, bioethicists and theologians, biotech leaders, and other seasoned activists about how to promote their cause on the local, state, national, and now international level. I was at the groundbreaking conference, where I first met Bernie Siegel face-to-face. I'm sure the opposition would like to characterize the activists at the meeting as baby-killing monsters (and they frequently do), but these people were ordinary Americans whose lives had been fractured by the random strike of some terrible disease or catastrophic injury. However, the energy and hope generated by the hundreds of activists at that meeting, some of whom were laboring away virtually alone in their communities, was electrifying. Many of them were starting to learn for the first time how individuals can organize, take action, and make their voices heard.

Since then, the Genetics Policy Institute has become involved in many more initiatives to promote public policy that supports stem cell research. Bernie Siegel is now an internationally sought-after speaker because of his unique knowledge of the U.N. and his experience in bringing diverse groups of people together. The network of individuals and activists that GPI engages to help stop anti-research legislation continues to grow, along with the number of state initiatives aimed at promoting stem cell research. Only recently, Bernie remarked to me that the unlikely victory in 2003, qualified though it was, followed by that first meeting in Berkeley, "felt like the beginning of a new civil rights movement."

As for the Costa Rican treaty at the U.N., in October 2004, the General Assembly decided to delay the vote once again.[6] The battle raged on until the fall of 2005, when finally, the Bush administration abandoned the hope of a worldwide treaty banning both types of cloning. Instead, they called for a nonbinding declaration, calling for all nations to desist from allowing "human cloning" whenever it is "incompatible with human dignity." In a sparsely attended session of the General Assembly, with fewer than half of the delegates voting, the declaration was approved.[7] This watered-down statement was ambiguous enough for *both* sides to claim victory, but to the relief of the scientific and patient communities, it meant that for now at least, the threat of a worldwide ban on therapeutic cloning was defeated. However, the battle to keep research cloning legal at the national level continues in the United States and in several other countries as well. Ironically, as of this writing, there is still no international law to ban human reproductive cloning. Patients are still fighting for their lives, and those who want to ban the research still believe they are fighting for God. Like an irresistible force meeting an impenetrable object, it isn't likely that either side will back down any time soon. In the next chapter, I'll show how, in Washington, DC, these issues have been exploited in a real-life political theatre of the absurd.

chapter six

political spin and the weapons of mass distraction

I just don't see how we can turn our backs on this. We have lost so much time already. I just really can't bear to lose any more.

—Nancy Reagan

Before long, we'll be harvesting body parts from fully formed people.

—Pat Robertson, televangelist

Politics is more difficult than physics.

—Albert Einstein

Those of us who are lucky enough to be able-bodied can't imagine the struggles encountered every day by the paralyzed. For Susan Fajt, a paraplegic, something as simple as a pebble on the sidewalk or a six-inch curb can be transformed into a major obstacle. Fajt has been in a wheelchair since 2001, when she was a passenger in a car driven by her then-fiancé (who I'll call Steve), on a stretch of road in Oklahoma. Steve lost control of the car, which flipped at high speed, resulting in a crash that, by Susan's account, should have killed them both.

Instead, Susan was pulled out of the wreck with a broken jaw, some fractured ribs, multiple cuts from the glass of the shattered windshield, and her back broken in two places. She doesn't remember being rescued by paramedics; she has only the nightmarish memory of waking up in the hospital trapped in a body with no feeling or movement below the chest. In the beginning, it seemed that Steve was in even worse shape. A severe brain injury had plunged him deep into a coma, where he remained for the next two months. When he awoke, he couldn't walk, talk, or do anything to care for himself. For Susan, it took a while for the double shock to sink in that she had lost both her own life as she knew it, and possibly also the man she loved. The beautiful, intelligent 24-year-old went from having a bright future laid out before her to trying to piece together a life that had been shattered into a million pieces.

As dire as her situation was, Susan was more fortunate than many people who find themselves suddenly paralyzed by a spinal cord injury. She had a supportive family with the means to seek out every available therapy that might help her. Even so, her alternatives were extremely limited, and none of them offered her the hope of ever walking away from the wheelchair that was now her only means of mobility.

Although her doctors gave her no hope that she would ever walk again, they told her that physical rehabilitation would be critical in order for her to achieve the highest level of functioning possible given the nature of her injury. If she had any capacity for regaining any movement or sensation below the chest, the therapy would maximize her chances. A range of exercises would help strengthen parts of her body, such as her arms and shoulders, which she would have to rely more heavily on. Having the immobilized parts of her body moved and stretched by therapists would help to stave off muscle contractions and advanced atrophy. And she would have to push herself mercilessly to learn to do things, like standing, that she formerly never gave a thought to. Once she was ready to begin rehabilitation, she searched out the best centers and ended up at a Houston clinic that was rated as one of the best in the country.

The therapists in Houston started her on the standard regimen for paraplegics, but Susan wasn't ready to resign herself to measuring improvement in tiny increments; what she wanted was to walk again, even if it was with assistance. The doctor there told her that she would never be able to walk, even with leg braces, but a part of her just couldn't accept this depressing prognosis. She kept looking for a better answer. She didn't expect to get up and run a marathon, but she desperately wanted to overcome at least some of her total dependence on the wheelchair.

Susan searched intrepidly for help. Over the next two and a half years, she continued to try various therapies, visiting one clinic after another in search of a program that would help her. She also searched out, and tried, less conventional treatments like massage therapy and acupuncture. But the small improvements always came in baby steps and were agonizingly slow to materialize. By 2003, her relationship with Steve had ended, and she was living in Austin, Texas, with her parents. As fate would have it, Steve's condition had improved dramatically. He was able to walk, to talk, and to live with a minimum of assistance, while Susan remained confined to a wheelchair.

One day, while searching the Internet, she learned about an experimental treatment being given to spinal cord patients by Dr. Carlos Lima at Hospital Egaz Moniz in Lisbon, Portugal. The treatment, which was not available in the United States, was said to involve harvesting adult stem cells from the patient's nose and transplanting them into their spinal cords at the site of their injury. Dr. Lima, a neuropathologist, claimed that the stem cells found in the olfactory tissue formed new nerve cells that "patched over" lesions in the spine. He didn't promise a complete cure for paralysis, but the procedure, which had been performed on only a few patients, sounded more promising than any alternative that Susan could find in the United States or anywhere else.

Intrigued by this development, Susan read about patients who had received Dr. Lima's treatment. Some of them reported that they had regained some sensation and muscle control below the sites of their spinal cord injuries, something that none of Susan's American doctors even thought possible. She thought it over. She knew she was taking a risk having a procedure done in a foreign country where the research protocols might be far less strict than in the United States. But if all she got from the treatment was voluntary bladder control, it would mean a lot to her. So she became the third American to fly to Portugal to try the new procedure.

Today, she recalls how outdated and ill-equipped the hospital in Lisbon was by American standards. However, when Dr. Lima sat down and talked with Susan in his office, he seemed confident that his treatment would take her well beyond the point that any of the traditional therapies would take her. He seemed to really believe not only that his experimental therapy could patch up injured spinal cords, but that he would have Susan walking again. However, he told her that the healing process took a long time and that she may not see any improvement for up to two years following the surgery. During that time, she would have to continue her physical rehabilitation therapy nonstop. It would be a long road, but Susan was hopeful. She decided it was worth a try.

Fajt underwent two surgeries at the hospital in Portugal. The first one removed about a quarter of her olfactory tissue, which covers an area about one inch long in the upper nasal cavity. This specialized tissue, which enables us to smell, contains a mix of different cell types, and some are thought to be neural stem cells, which give rise to neurons and other nerve cells. The olfactory tissue is believed to have greater-than-average regenerative ability. Its progenitor cells (the stem cells), in addition to stimulating the birth of new nerve cells, are thought to secrete neuronal growth factors that can help to heal damaged nerve tissue like the damaged tissue in Susan's spinal cord. Still other cells residing in this tissue are remyelinating olfactory ensheathing cells, which insulate nerve fibers and facilitate communications between nerve cells. Dr. Lima's theory is that, after being transplanted into the patient's body, the tissue can create a close proximity to spinal cord tissue, with its numerous cell types, with no danger of rejection because it comes from the patient herself.

After Susan's olfactory tissue was removed, Dr. Lima sliced it into tiny pieces. He didn't attempt to isolate the cell types or multiply the stem cells in culture (a step that is generally used in adult stem cell transplants). Instead, he hoped that the chopped-up tissue would take its cues from the surrounding tissue in Susan's spinal cord and provide all the necessary cell types to create functioning nerve tissue.

The second surgery was extremely risky. It involved actually cutting into the spinal cord at the injury sites. Any mistakes or slips of the scalpel at this stage could have devastating and permanent consequences. Dr. Lima removed as much of the scar tissue from the old injuries as possible, and made small holes or cavities where the scar tissue couldn't be removed. The tiny pieces of olfactory tissue were implanted into the cavities created in the spinal cord. The membrane covering the spinal cord was closed up, the incision in Susan's back was sewn up, and the surgery was finished. Soon she was on her way back to the United States, to keep up a grueling physical therapy regimen and to wait for new, functional tissue to grow in her spinal cord.

The treatment Susan received was not cheap and it was not covered by insurance. She paid a Detroit clinic that coordinated the process $75,000 out of her own pocket. But most crushing of all is the fact that almost three years after the surgeries and continued physical therapy, her progress has been far less than what she expected.

There has been some improvement. She has regained some of the sensation in her lower body, has better bladder control, and has some voluntary movement of the muscles in her thighs. With the help of leg braces and a walker, she can now stand for short periods and even take a few steps. But in her day-to-day life, she is still completely dependent on a wheelchair. It's also impossible to sort out at this point to what extent, if at all, the cellular therapy performed by Dr. Lima is responsible for her modest improvement. Some of her improvement could be due to the removal of the scar tissue and the resulting decompression of her spinal cord. And without a doubt, her youth and her unflagging dedication to the rigors of physical therapy have also played a role. She still works out for two to three hours a day on a machine that she and her father invented just for paraplegics, to maximize the benefits of therapy.

But in spite of the gains she has made, the cellular therapy has far from lived up to her expectations. She no longer believes that the cells Dr. Lima transplanted into her spine were stem cells, because the expected regrowth of nerve tissue does not appear to have happened, or at least it happened on such a tiny scale that it has had little functional effect. "I'm still paralyzed," she told me when I spoke with her in late 2005. Of the treatment she received in Portugal, she told me, "In no way, shape or form is this a cure for paralysis."

Susan has since talked with other patients who were treated by Dr. Lima and found that their improvements were also minimal, and hard to pinpoint to the cellular transplants. Because Dr. Lima has never published his results in any science journal, it's hard to draw any conclusions about whether his patients have even acquired new nerve tissue or how functional that tissue is. All anyone has to go on is the anecdotal experiences of the patients, but so far no one seems to have been miraculously cured of paralysis. Susan now thinks that the clinical trials in Portugal should be halted until someone comes up with a better protocol and better results to justify the cost to the patients. "People are selling their houses to go to Portugal to do this," she says. "And it isn't worth it."

Susan's journey didn't end with a disappointing attempt at a cure that left her $75,000 poorer—far from it. To her dismay, the fact that she traveled overseas to receive the experimental adult stem cell treatment has made her into a kind of poster child for many right-to-life groups, who have exploited her story in articles and postings all over the Internet. These articles grossly exaggerate the benefits she received from the transplant in order to buttress their case that adult stem cells can replace embryonic ones. But in her opinion, the most exploitative treatment of all came from a conservative congressman from Florida.

In July 2004, one year after her treatment in Portugal, Susan was asked to testify at a Senate hearing on stem cell research in Washington, DC. At the hearing, which was chaired by Senator Sam Brownback of Kansas, she told a roomful of senators about her treatment and credited some of her improvement to the treatment she'd received in Portugal. She also made it clear that she was a long way from being cured. She made a plea for increased funding of all kinds of spinal cord research, and for the government to subsidize rehabilitation services for the paralyzed. Fajt, who is in favor of all kinds of stem cell research, had no idea that her appearance at the hearing meant that her experimental treatment was going to be hailed as an astonishing cure and used as an argument against federal funding for embryonic stem cell research.

Immediately after the hearing, a woman approached Susan and asked if she could take a photograph of her next to a man she didn't recognize. The man happened to be Dave Weldon, a Republican Congressman from Florida. Not only did this woman want a picture, but she and Weldon insisted on having Susan stand up for it, which Susan thought was a pushy and inconsiderate request. It's no easy feat for her to be lifted out of her wheelchair and stand, even with leg braces and a walker. At first she declined. "It just wasn't the time or the place for it," she told me when I talked with her, "but they absolutely insisted that I stand up for the picture." Finally, she gave in, but something about the situation didn't feel right. Susan had this to say about Congressman Weldon: "He was very cold. I did not like that man. Something was definitely wrong—now I know he was using me to harm my cause."

The last impression was confirmed later, and only by chance. In May 2005, Fajt turned on the evening news from her home in Austin. A story came on about the debates in Congress concerning the passage of HR 810—the bill to modestly expand federal funding of embryonic stem cell research. To her disbelief, Fajt saw Congressman Weldon holding up a poster-sized photo of her "standing" next to him. The bottom of the picture was so dark that her leg braces vanished into the shadows. Anyone looking at the photo could easily believe Susan was standing and perhaps walking with no more assistance than that of a walker. "When I saw what he was doing, I almost fell out of my wheelchair," she said. Even worse, Weldon was also using her story as highly misconstrued "evidence" that adult stem cells are already curing patients like her. Using Susan's name and face, he told the lawmakers in Congress, as they prepared to vote on the stem cell research bill, that embryonic stem cell research is both unethical and unnecessary. "He had used my image, without my permission, to distort everything I believed and fought for," Susan told me. "He didn't just hurt me, he hurt every person who could benefit from the research." She was furious.

She immediately wrote a letter to Weldon demanding that he stop using her image and asking for an apology, and e-mailed it to him. Weldon stopped using her photograph, but he never apologized and never attempted to correct the false impression he had made on the floor of the House of Representatives that Susan was far more ambulatory than she is.

As of this writing, he still has a highly misleading story about Susan posted on his website, titled, "Why Embryo Stem Cell Research is NOT the Answer." The text refers to Susan as "previously confined to a wheelchair."

In reality, Susan is very much confined to a wheelchair, and is frustrated that politics are hobbling the science. She wants to see the experiments in which embryonic stem cells were used to reverse paralysis in rats carried over into humans, and to see this done in the United States so that American patients don't have to go overseas to receive treatments. "It's so scary that our own government denies us a cure," she says of the Bush policy. "They've reversed paralysis in rats, but I'm afraid we're going to be walking on Mars before people like me walk here on earth. All I want to do is put my footprints in the sand again. How can these politicians allow needless suffering because they care more about cells in a dish than they do about people like me?"

Since Susan's brush with the cynical politics of the far right wing, the political grandstanding over stem cell research has become even more extreme. One strategy of the anti-embryonic stem cell activists has been to try to convince the public that adult stem cells are a medical panacea, as was done in Susan's case. Another tactic has been to cast embryonic stem cell research in the most diabolical, and baroquely embellished, light imaginable.

In May 2005, Senator Sam Brownback reintroduced in the Senate his Human Cloning Prohibition Act, the infamous bill I highlighted in chapter four. He held a press conference the same day in the Senate Dirksen Building, one of the enormous, labyrinthine buildings that form a cluster in the heart of Washington, DC, known as Capitol Hill. I attended the conference in order to stay abreast of the latest legislative developments that could affect stem cell research.

When I slipped into the room where the press conference was being held, it had already started. Mr. Brownback, a pale, young-looking senator with impossibly neat, combed-over brown hair, was standing at the front of the room behind a large podium. The bill's sponsors in the House of Representatives, Congressmen Dave Weldon and Bart Stupak from Michigan, were standing beside him, to show their support of the bill. Also there to lend their support for the Human Cloning Prohibition Act was a group of representatives from both ends of the bioconservative spectrum. The seemingly ubiquitous David Prentice from the Family Research Council was there, plus representatives from the National Right to Life Committee, a liberal environmental group called the International Center for Technology Assessment, a left-wing anti-biotechnology organization called Friends of the Earth, and one paralyzed patient in a wheelchair. Senator Brownback, ever the strategist, had chosen these speakers to create the illusion that support for his anti-cloning bill covers a broad political spectrum, when in fact it is supported only by the *extremes* at either ends of the spectrum. There are a few liberal groups that do support anti-stem cell research legislation in general, but they are few and far in between; the vast majority of anti-research groups are from the far right end of the continuum.

The event was clearly scripted from beginning to end. Senator Brownback announced the reasons why he was reintroducing the bill (which must be reintroduced each year if no action was taken on it the previous year). The bill's supporters took turns at the podium, speaking about the dangers of therapeutic cloning and why their organizations want it criminalized. I can only describe these statements as bizarre. One by one, they painted a picture of an American scientific community so dark and dystopian that anyone who missed the introductions might think they had stumbled into a science fiction writer's workshop—and a markedly twisted one at that.

As I sat there listening to what these anti-stem cell research activists had to say, it struck me as utterly surreal that such an imaginative performance could even transpire within such an official U.S. government setting. What was being presented to the press were bold-faced fictions, but through the sheer power of repetition, their purveyors hoped that these myths would be believed by presumably gullible media and an unsuspecting public.

"There is no difference between therapeutic and reproductive cloning," was one often repeated statement. "Embryonic stem cell research has had a complete lack of results with animals. Most of the experiments have failed," was another one. "Placental stem cells and umbilical cord stem cells are already successfully treating paralysis and Parkinson's disease." "Seventeen patients have now been cured of paralysis with umbilical cord cells." "If we allow embryo cloning, it's just the first step. Scientists will inevitably start to do in people what they are already doing in animals—letting the embryo grow into a fetus so they can harvest the organs." As the scientific research shows us, the claims here are grossly inaccurate, but the speculation didn't stop there.

"Since it has no hope of curing patients, we have to question the motivations of those scientists who want to do this cloning," said one of the speakers. We didn't have to wonder about the alleged motivations of American scientists for long, because presently we were told, "Cloning entails deliberately creating human beings in order to use their body parts." These weren't just wild slips of the tongue. One of the print handouts at the conference, from the group Concerned Women for America, an organization whose stated mission is to bring biblical values to public policy, railed, "Once scientists get approval for creating, experimenting [sic] and killing the smallest of cloned humans, their incessant push for no moral boundaries will extend past the embryo stage to cloned fetuses (unborn babies), then onto newborns and beyond."

As time went by, the claims became more shocking and outrageous, and the image of American scientists entered cartoon-villain territory. However, the small group of reporters (who presumably should have been at the riot stage over what they were hearing) sat slumped in their chairs, quietly taking notes. Finally, one of the speakers delivered his *pièce de résistance:* "This is the first time in history," said the speaker, "where we would force women to have abortions, if we're going to fulfill the law."

It would be easy to laugh at such a performance if it weren't so sad that this shameless theatre has come to characterize our national political life. Senator Brownback and his friends have continued, at every opportunity, to paint a picture of pure science fiction for the voting public. That picture is of a dark, lawless world where B-movie-style evil scientists lurk around every corner, insatiable in their desire to create, experiment on, and torture small children. They want us to believe that we stand on the brink of a state of moral pandemonium, where soon, a diabolical U.S. government will force moms to abort their babies to feed the soulless machinery of scientific exploitation, while patients are denied cures using adult stem cells simply because scientists are only interested in cures that kill innocent babies.

I thought that Senator Brownback's political sideshow was about as extreme as anything I could imagine, but there were more sensational developments in store during the summer of 2005. Conservatives in the Senate began discussing at least four new anti-embryonic stem cell bills, which they are calling "alternatives" bills. These bills propose to devote government funding specifically for research projects that provide "alternatives" to the use of embryonic stem cells. The only problem is the alternatives being proposed range from the scientifically unproven— "reprogram adult cells and return them to an embryonic state"—to the confusing, yet deeply unsettling—"use only dead embryos" and "use only embryos taken from dead people." This last suggestion was actually made by a member of Congress to Congresswoman Diana DeGette, as he tried to talk her out of introducing HR 810. One has to wonder if a member of Congress actually thinks doctors should start examining dead women to search for possible embryos that may chance to be sitting in their fallopian tubes!

Even though alternatives bills are being pushed as solutions, in reality there is absolutely no need for them—there is no restriction on grants to fund legitimate research aimed at developing alternative sources of stem cells. As long as the research has merit and can pass the NIH's peer-review process, which is how proposed research projects are approved for funding, there's no obstacle to obtaining funding. This is amply illustrated by the NIH's expenditure of over $500 million per year on animal stem cell and human adult stem cell research, much of which is already being used in the quest to reprogram adult cells into an embryonic state. So where's the problem? The alternatives bills, by seeking to fill a void that doesn't exist, can be nothing more than political posturing on the part of politicians who are desperate to appear pro-research while still pacifying their extreme right-to-life constituencies. This way, senators and congressmen will be able to point to their votes for these meaningless bills and say to voters, "See—I'm not anti-stem cell research because I voted for one of these 'pro-research' bills."

One of my strangest brushes with politics since I began making visits to Capitol Hill is an incident that happened in the summer of 2005 during a visit to Virginia Senator George Allen's office. I went there with two other members of the Coalition for the Advancement of Medical Research to brief the senator's legislative aide on why the medical research and patient communities want real pro-stem cell research legislation to pass.

We were welcomed by the aide, a very courteous young man who appeared to be in his late twenties, who walked us to an inner room with a long conference table. As we seated ourselves, he told us he had just spent all day in meetings with right-to-life groups, hearing their side of the story on the "alternatives" bills. He seemed shell-shocked by the meetings, and before we even got started, he wanted to impress upon us that Senator Allen was especially concerned about the need to pass a new anti-chimera bill, another bill introduced by Senator Brownback.

The word "chimera" comes from the name of a fire-breathing monster in Greek mythology. In mythology, it's a freakish creature with a lion's head, a goat's body, and the tail of a serpent. In science, however, the term chimera has a far less spectacular meaning. It simply means an organism that harbors more than one set of genes. Most cases are research animals that have received human cells or snippets of DNA to induce them to develop an animal version of a human disease. A good example is a breed of mice that has been developed with small amounts of human DNA that give it a mouse version of Alzheimer's disease. Scientists are now using these mice to test new therapies for Alzheimer's disease. But more to the point, researchers have started to transplant human stem cells into animals to see how they work in a living system. The presence of human cells (and human DNA) in these animals makes them chimeras. This is seen as a critical step in the research because scientists can't safely go straight from testing cells in a petri dish to implanting them into people. Chimeras allow scientists to test for possible adverse effects before transplanting the cells into humans. But the right-to-life community is now taking up the crusade to stop researchers from being able to do this. Senator Allen's aide told us that he was still reeling from his last meeting, where he had heard some "very disturbing things."

As we sat around the table, preparing to verse him on the need for scientists to be able to do therapeutic cloning, he suddenly asked us, "What do you call those creatures on the evolutionary chart that aren't human—they're still apes, but they've just started to stand and walk upright?" This startling question, which seemed to come from straight out of the blue, hung in the air for several seconds, before I ventured, "Oh, you mean *homo erectus*?" "Right," he said. "Well, we don't want any of those."

Again there was a stunned silence. No one knew what to make of this. I even wondered if he was joking, but he brought the issue up two more times, with perfect seriousness.

For the rest of the day, I couldn't get this conversation out of my head. I called Bernie Siegel to tell him about the strange comment, and the first words out of his mouth were "Planet of the Apes!" Of course. As outlandish as it was, I realized that this was exactly what the aide was talking about. Some lobbying group had gotten to this earnest young man and convinced him that stem cell scientists are careening down the side of a greased mountain, on their way to the creation of a subhuman race of ape-men (presumably to serve their human overlords, no less).

All fictions aside, what is one to think about the terrifying scenarios of scientists harvesting organs from fetuses and even babies, and trying to create monstrous creature hybrids? Scientists say they are trying to cure disease, while some politicians and political activists claim they are the purveyors of absolute evil. How could there be such a radical contrast between what scientists are saying they want to do and the claims of politicians and right-to-life activists? In chapter three, I provided a snapshot of what the real research consists of, along with the thoughts of some of the scientists who are doing it. Many people wonder how it has come to be that the actual science has been largely drowned out by the political spin that opponents have applied to it.

To some extent, the situation can be more easily understood if we take into consideration the vast cultural differences between scientists and politicians, and how these differences have contributed to the crippling stalemate we now find ourselves in. A lot of the problems in public perception of the research can be attributed to the immense disparity in the language of science and the language of politics. Scientists tend to express themselves in an understated way, a style which is ingrained in scientific culture, where the prevailing attitude is one of trained skepticism. If a scientist overreaches and makes unfounded claims about her research, the mistake is very likely to be discovered by other scientists, and her loss of credibility could be devastating. When a researcher is seen by her peers as being less than credible, not only is it an embarrassment, but the loss of reputation can end a promising career and derail many years of hard work.

We should also keep in mind that science is a mode of inquiry, not set of scriptures or even an encyclopedia of absolute truths. Even accepted facts will be tested again and again. The conclusions of any research effort are not accepted as fact until they have been duplicated and verified by others. As Einstein once said so eloquently—if scientists knew what they were doing, they wouldn't call it research. Generally speaking, while it is far from true that scientists are incapable of making false or exaggerated claims, science is a self-correcting enterprise. When your work is being published and scrutinized by your colleagues (and competitors), it is very likely that mistakes will be discovered. When scientists are interviewed by journalists, it behooves them to express themselves very carefully, sticking only to the facts. But this conservative communications style doesn't grab many headlines in a media environment that often focuses on sensationalism. The careful, reasoned style that works with the peer-review boards of scientific journals can easily be drowned out by the more splashy tones of the nightly TV news and the local newspaper.

The language of political activism, on the other hand, especially among today's right- and left-wing extremists, is often red in tooth and claw. A verbal campaign of shock and awe can make or break a political campaign. It *begins* with unabashed dogma and proceeds to exaggeration, hyperbole, and drama. Unfortunately, emotionalism is the currency of today's politics. Getting the facts right takes a backseat to inciting a strong emotional reaction that will grab headlines and garner votes. Political extremists rarely acknowledge nuances, and there are certainly no shades of gray in the stark, black-and-white world of pro-life activists and left-wing extremists. The only things that matter are the absolutes, because, in the end, they are often what motivate people to take action.

No one understands this better than Sam Brownback. The content of his Human Cloning Prohibition Act is appalling to most Americans when they understand what he's really proposing. The senator probably knows that he can count on his social-conservative base to support him, but he also needs to get the bill approved by more moderate Republicans and the Democrats in the Senate. If he can overwhelm people's minds with terrifying images of dismembered babies, hopefully that will drown out the fact that his bill would put doctors and patients in jail and places a higher priority on microscopic cells in a lab dish than on suffering patients.

Anti-research activists have been claiming for several years now that there are *scientific disagreements* about the relative potential of adult stem cells versus embryonic stem cells. This is simply not true. The disagreement is not among scientists about what lines of stem cell research have the most potential. The vast majority of scientists see both types of research as complementary and believe that research into all types of cells should be pursued. The disagreement is really one of political and religious ideology. It is between those who believe that *all* kinds of stem cell research, including embryonic stem cell research, should be pursued, and those who think that *only* adult stem cell research should be conducted.

Many Americans have been confused by those who object to embryonic stem cell research on moral and ethical grounds professing, as Congressman Weldon does, that there are scientific reasons not to pursue it. A handful of scientists, such as Prentice at the Family Research Council and Jean Paduzzi, of the University of Alabama and the right-to-life organization Do No Harm, are clearly individuals who object to embryonic stem cell research for ethical reasons. They are regularly pitched to media outlets and appear at government hearings as "concerned scientists," but make no mistake about it: They perform this role in the service of right-to-life political action groups.

So what about the aforementioned "alternatives" bills that have been proposed by U.S. senators as possible ways to sidestep the use of embryos as sources of stem cells? On June 30, 2005, the *Washington Post* ran an article titled, "GOP Probes Non-Destructive Stem Cell Research," which outlined the proposed alternatives. Financial support for scientists willing to research these so-called alternatives has been proposed by Senate Majority Leader Bill Frist, Senator Rick Santorum, a far-right Republican from Pennsylvania, and several other conservative senators. The theories that these politicians have decided need support are based upon some recommendations that were published in a paper released in May 2005 by President Bush's council on bioethics (an advisory body that has been criticized for its ideological tilt toward the extreme right) and supported by the president himself.

In the *Post* article, Senator Frist said, "All of the research you have there [in the four alternatives] stops short of the creation of an embryo for experimental purposes, and short of the destruction of an embryo for experimental purposes. That is the direction I think we should explore." The four suggested alternatives for deriving embryonic stem cells are "altered nuclear transfer," "nonlethal biopsy," obtaining stem cells from technically dead embryos, and inducing adult cells to become the pluripotent "master" cells. All of these theories have some critical scientific stumbling blocks, and some of them may not even avoid the ethical objections of those who oppose embryonic research.

One of the most well-known proposals, offered last year by Dr. William Hurlbut, a physician as well as a member of the president's bioethics council, has been named altered nuclear transfer (or ANT, as he calls it). This involves using nuclear transfer (or therapeutic cloning) to create what might be called a "developmentally scrambled" embryo. The DNA of the donated adult cell would be altered so that a key developmental gene—the "decision-making" gene that drives the organization of the cells and tissues to create an organism—would be disabled. This critically altered DNA would be inserted into an egg, and rather than creating a normal embryo, it would create a chaotic mass of human cells and tissues, from which, (theoretically) an embryonic stem cell line could be derived. Such masses sometimes occur in the human body in the form of tumors called teratomas. Teratomas are ghastly, disorganized masses of cells and tissues, where teeth, bone tissue, and hair can mix with soft organ tissues in a cancerous growth.

It's somewhat surprising that this theory has received so much attention, considering that it has at least two obvious problems, one scientific and one ethical. From a scientific standpoint, the idea of someday transplanting cells lacking such a critical gene into the human body is concerning, to say the least. Genes seldom serve a single purpose and are part of a complex interplay involving other genes and even the environment. It's risky to assume that cells created in such an abnormal way would be normal, or would divide and develop normally once they were transplanted into the body.

On the other hand, it's not at all clear that Hurlbut's plan would be free of ethical objections. Some critics would point to the fact that ANT goes through the motions of human cloning while critically "injuring" an embryo that, without the last step, could have potentially become a human being. Those who see a potential human in every cell with a complete set of chromosomes might see this as the deliberate creation of a "human" monster, and the "yuck factor" alone turns many people off. Would you want the cells of a teratoma, which don't know whether to become heart, pancreas, bone, teeth, or hair, transplanted into one of your vital organs? Even the conservative syndicated columnist Charles Krauthammer has called the technique "repugnant and weird," while maintaining, as many have, that ANT still creates an embryo, and a fatally injured one at that.

Nonlethal biopsy is another proposed alternative that has received attention over the last year or so. It involves removing a single cell from a very early-stage embryo (a ball of about eight cells) and using that single cell to create an entire self-replicating line of pluripotent stem cells. Although no one has done it, it could some day work. Removing one cell at this stage, which is routinely done at in-vitro fertilization clinics, doesn't destroy the embryo, which can still grow into a normal baby. However, this approach may not be as ethically neutral as it seems at first glance. It is at least possible that that one cell could also develop into a complete individual.

As I mentioned earlier, embryologists working with animals have been able to extract a single cell from the embryos of several species, and use that single cell to grow an entire organism—in effect, creating genetically identical twins or triplets, much as nature would, from a single embryo. This suggests that a single human embryonic cell, separated at just the right time and under the right circumstances, may have similar potential. If it does, then logically and ethically it would place the single embryonic cell into a similar category as the fertilized egg. Nonlethal biopsy is not certain to quell the objections of those who think that any and all human potential is sacred. In fact, as science continues to push the frontiers of what is possible in assisted reproduction, our boundaries for what we consider to be human potential are bound to keep shifting, and to be the subject of ongoing disagreement.

One of the most unusual and unbelievable theories to come out of the alternatives agenda is the plan to derive living cell lines from technically dead embryos. For scientists, the definition of embryonic death begins when the microscopic cells fail to divide. It's difficult to imagine how cells that have lost the ability to divide in-vitro could ever be revived to the point that they would begin dividing with the energetic potential of embryonic stem cells. If there are scientists who want to pursue this idea, there is nothing to prevent them from doing so. However, the effort to derive living cells from dead ones doesn't sound like a good use of our research dollars, especially when funding is being denied for research that we already *know* produces pluripotent stem cells.

The fourth alternative, and one that is already being avidly pursued by scientists throughout the world, is to find a way to reprogram adult cells to return them to a state of pluripotency. Despite repeated claims on the part of right-to-life activists that such a feat is already being performed, so far the efforts to turn adult cells into normal, therapeutically useful embryonic cells have been unsuccessful.

A recent development in this ongoing effort came in August 2005, when several highly respected Harvard scientists announced that they had succeeded in reprogramming adult skin cells back to a pluripotent state. The research team, which included Kevin Eggan and Douglas Melton, a prominent embryonic stem cell researcher, involved using chemicals to fuse adult skin cells with embryonic stem cells. The result was a hybrid cell that contained the DNA of the skin cell donor, much as a cloned cell would, and the cells were determined to be pluripotent. However, the hybrid cells contained exactly double the normal number of chromosomes. For the cells to be suitable for medical treatments, someone would have to figure out how to extract a complete set of chromosomes from them, and so far no one knows how to do that. The cell fusion technique is also extremely inefficient. The Harvard scientists found that it took about 50 million skin cells and 50 million embryonic stem cells to produce just ten or twenty hybrid cells.

Because of the cells' genetic abnormality, Dr. Eggan emphasized that the breakthrough doesn't offer an alternative to embryonic stem cells. What it *does* offer is an experimental tool that may help scientists understand how adult cells might be reprogrammed. And once again, this particular technique doesn't get around the necessity of using embryos in research, since it required embryos to obtain the pluripotent cells used to fuse with the adult cells in the first place. This development on the cellular reprogramming front is just a reminder that scientific research is a long and painstaking process. As much as we'd like to believe there are shortcuts, the secrets of cellular programming could take many years to unravel.

The difficulty is that terminally differentiated adult cells—say a skin, blood, or kidney cell taken from the fully developed human body—are profoundly different from embryonic stem cells. The sought-after quality of pluripotency depends on a myriad of factors, including the switching on and off of certain genes, the assembly of proteins and other molecules, and another crucial ingredient—interaction with factors in the environment of the egg. It is entirely possible—probable even—that unidentified factors existing only in the living oocyte (or egg) are utterly indispensable to triggering the development of normal pluripotent stem cells. Once a cell has completed the journey to becoming a muscle, bone, or liver cell, its embryonic properties, plus the conditions that triggered them, no longer exist. Again we confront the familiar problem: We may never be able to identify, duplicate, or control these factors without the study of human embryos. No matter how much research is conducted on adult cells, it is highly doubtful that they will ever reveal the secrets of pluripotency or tell us how to capture the proliferative ability of embryonic stem cells.

Sooner or later we have to ask if it will ever be possible to pursue the extraordinary promise of embryonic stem cell research while at the same time satisfying moral absolutists. While scientists and ethicists go through increasingly complex contortions in trying to reconcile science with fundamentalist religious beliefs, the exercise may be analogous to chasing an elusive horizon—the closer you get to the horizon, the farther it moves away from you. A perfect illustration of the conundrum comes from Dr. Michael Gazzaniga, a member of the President's Council on Bioethics, and a dissenting voice in the council's report recommending research into the four alternatives.

Gazzaniga has pointed out that even the reprogramming of adult cells to return them to an embryonic state will not satisfy absolutists. His reasoning points to the same conundrum I mentioned earlier—if adult cells are somehow sent "back in time," developmentally speaking, won't there be cries for parents to "adopt" them and laws to protect them? And won't these cells, being genetically identical to the cell donors, in fact be just another way of cloning?

Finding a true alternative to research using embryos is certainly a worthwhile goal that would free this beleaguered field of much of its controversy. Those who oppose the destruction of embryos would have achieved their goal, and scientists would be happy because it would free up federal funding and take them out of the line of political fire. Today some of the field's most prominent scientists, including Rudolf Jaenisch and Harvard's George Daley, are trying the altered nuclear transfer technique in mouse embryos to see if it works. But to allocate special funding for research that serves political ends while banning research that has shown more promise but offends a highly vocal minority would turn our whole system on its head. To expect scientists to pursue lines of research based not on valid data or on their own best judgment, but on the desires of politicians, makes no rational sense and seems likely to lead to a waste of resources. In November 2005, Daley told the *New York Times,* "How many hoops do you have to go through as a scientist, when you don't think you are doing anything wrong?"

Michael Gazzaniga hit the nail on the head when he added his personal statement to the "alternatives" paper issued by the President's Council on Bioethics. He asked, "Why delay what we know works with this sideshow?" and finally, "Is the United States of America going to allow embryonic stem cell research and biomedical cloning to go forward using the now widely accepted techniques used by the private sector, by the State of California, and by dozens of other countries, or is it going to remain hostage to the arbitrary views of those with certain beliefs about the nature of life and its origins?"

Of course, if it were a matter of making such a simple, innocuous choice and still coming up with cures for many of mankind's most devastating conditions, it's hard to imagine anyone objecting to any one of these alternatives. But whether any of these theoretical alternatives will ever lead to cures is open to question and could take years to determine. It's no secret that many of the same vocal conservatives who want to promote the presumed alternatives are doing so as part of a political strategy to diminish, if not eliminate, government support for embryonic stem cell research. The alternatives agenda, rather than being a matter of exploring *all* the avenues to determine which is the most promising, includes closing off the one avenue that so far has shown the most promise, and lets politicians—not scientists—decide which lines of research receive public funding.

In the next chapter, I'll explore some of the ethical issues that underlie the explosive politics of embryonic stem cell research.

chapter seven

ethics and the embryo

The energy of the mind is the essence of life.

—Aristotle

As people of faith we are called to be partners with God in healing and in the alleviation of human pain and suffering. . . . With careful regulation, we affirm the use of stem cell tissue for research that may result in the restoring of health to those who are suffering from serious illness.

—Presbyterian Church, USA

The process kills the days-old unborn child.

—Steven Ertelt, LifeNews.com

A huge number of frozen embryos sit in tanks of liquid nitrogen at in-vitro fertilization clinics across the country. Held in test tubes that are suspended in the cylindrical metal tanks, in temperatures of minus 320 degrees Fahrenheit, their very existence has become a lightning rod for controversy in a field that was already the focus of impassioned disagreement. These embryos have become hostages in the war over stem cell research, stuck in a state of frozen limbo. Some people believe they are equivalent to living human beings, while others see them as cells that may or may not ever come close to a critical threshold we call personhood.

The idea that human life begins with a single lightning strike at conception has had a powerful influence on both politics and religion. Extraordinary efforts are being made by conservative politicians at every level of government to formalize the status of the embryo as a full-fledged human being. Many believe that an embryo, from the moment of conception, should be granted the same legal protections as any man, woman, or child. At the same time, scientists, patients, and a majority of Americans believe it is acceptable—even desirable—to use very early-stage embryos in research aimed at the cure of disease.

At the very center of this dispute are the 400,000 frozen embryos. These blastocyst-stage embryos are an inevitable byproduct of the IVF process. With current-day science, there could be no in-vitro fertilization successes without producing large numbers of embryos, because for every embryo that results in pregnancy, there are usually several others that are either unsuitable for uterine transfer or that simply don't survive in the womb.

There is ferocious disagreement over what should be done with the blastocysts that are left over after a couple has either fulfilled its child-bearing goals or given up on the hope of a successful pregnancy. Some right-to-life groups say that fertility clinics should severely limit the number of embryos they produce so that there will be fewer left over. But even if we stopped creating embryos today, we would still be confronted with the question of what to do with the hundreds of thousands that already exist. Scientists and patient advocates say that those not slated for reproductive purposes should be used to derive embryonic stem cells for biomedical research. Others say it is acceptable to use leftover IVF embryos in research, but wrong to create cloned embryos to be used as sources of patient-matched stem cells. Never before has so much attention been poured into the question of the status of the early-stage embryo, and what can or should be done with it.

"Life begins at conception." The slogan has become so ubiquitous among anti-abortion activists that it's hard to believe that the idea of one celled personhood (a zygote at the moment of fertilization) doesn't come from the Bible. It doesn't. The concept was first proclaimed in 1869 by Pope Pius IX, a conservative pope who became known for his opposition to what he saw as the creeping liberalism among Catholics in the 1800s.[1]

Before the Vatican's pronouncement that abortion at any stage is considered murder, the Catholic faith had adhered mainly to the views of St. Augustine, who believed that the soul enters a fetus 40 days after conception. It is this "ensoulment" that marked the beginning of personhood, not the mere presence of human "matter." In the later years of the nineteenth century, the argument that life begins at conception was increasingly adopted by the American anti-abortion movement, but the idea only became widespread from the 1970s onward, when in-vitro fertilization entered the picture.

Complicating the debate about the acceptability of embryonic stem cell research is the fact that many people are confused about the differences between an embryo and a fetus. The two terms are sometimes used by opponents of stem cell research as if they were interchangeable, but there are important differences. According to *Dorland's Medical Dictionary*, the definition of an embryo is a developing organism beginning about two weeks *after conception*, until approximately six weeks of development. It's important to note here that the blastocyst-stage embryo, the embryos in question here, are regarded by many experts as "pre-embryos," rather than true embryos. If they were created by natural conception, they would still be in the woman's fallopian tubes, prior to implantation in the uterus. In addition, some experts don't regard a fertilized egg as an embryo until it attaches to the wall of a uterus and establishes a pregnancy. Technically speaking, the embryos being stored at IVF clinics that are possible sources of embryonic stem cells are pre-embryos.

A fetus, on the other hand, is at a later stage of development. The term "fetus" is generally applied seven to eight weeks after conception. Furthermore, an early-stage embryo can be artificially maintained for a few days outside the body while a fetus can only exist inside a woman's body. It certainly can't be maintained artificially, despite some of the more outrageous rhetoric about "growing fetuses for their spare parts." Although I've encountered alarmist statements about the possibility of "fetus farms," such a scenario is impossible, unless one were to regard the bodies of women in much the same way as we regard cows that are maintained to produce milk.

So what exactly is a blastocyst? Some right-to-life advocates today refer to the blastocyst as if it were a diminutive, fully formed human being, or a "homunculus." The homunculus idea has been around for centuries, and was most prevalent during the 1600s. The homunculus was theorized as a miniscule human being, complete in every way, that is folded up inside of sperm or egg cells. People believed for a time that these microscopic babies lived inside our bodies and were mysteriously activated in pregnancy to grow larger and larger, but without changing in any way. An early-stage embryo is definitely *not* a homunculus; there is not only a vast difference in size between an embryo and a baby, there's a vast developmental gulf as well.

Those who say that harvesting stem cells destroys a tiny person "for their body parts" are being grossly inaccurate. The blastocyst, which exists at about five to seven days after an egg has been fertilized, is a tiny, fluid-filled ball too small to be seen with the naked eye. It has an outer membrane of cells, and inside of this membrane, there is a tiny clump of about 8 to 200 undifferentiated, pluripotent stem cells. These cells are all identical "master" cells that have not yet begun to differentiate into any particular cell type. There can be no "body parts" because there is no body. A blastocyst cannot possibly feel pain because there is not a single nerve cell at this stage, never mind an entire nervous system or anything remotely resembling a brain. As Michael Kinsley, the renowned commentator who also has Parkinson's disease, has written, a blastocyst is made up of "a few dozen cells that together are too small to be seen without a microscope. It has no consciousness, no self-awareness, no ability to feel love or pain."[2]

For many people, the decision about whether embryonic stem cell research is permissible depends on when, in their opinion, human life—that is, the life of a *person*—begins. After all, the taking of a human life is something that we abhor, regardless of whether we can help another person by doing so. The future of embryonic stem cell research depends, quite simply, on whether we believe that a person is destroyed in the process. And the answer to that question cannot be provided by science—it is inevitably a matter of belief.

Science can tell us what an embryo is in physical terms, but no responsible scientist would claim to have "proof" of when cellular matter becomes a human life. The decision to equate an embryo with a full-fledged person cannot possibly be based on the embryo's form or function, its history, or even its relationships to others—that judgment is a purely religious and philosophical one. Religious views on the subject vary, but for most of us they provide a foundation for how we view the status of an early-stage embryo. As is reflected in public opinion polls about embryonic stem cell research (see chapter four), most of us think that a life that we consider *human* requires a more complex developmental foundation than that of the blastocyst, but there is by no means unanimous agreement on this subject.

On the other hand, it is only natural that we afford the human embryo special consideration, that we recognize and honor its potential. However, behind the rhetoric of the right-to-life movement lies an odd, unspoken assumption: that every embryo, if not destroyed by "murderous" scientists or abortionists, has a 100 percent chance of developing into a baby. We might think that nature conspires to afford each and every one of these special cell clusters with a maximal chance of survival. But nature operates by its own rules, often running counter to what we think should happen. And in nature, many more embryos are created than will ever make the journey to birth.

The science of embryology has recently disclosed some new information about the potential fates of embryos. In recent years, scientists have determined that even under the best possible circumstances, in a woman's body, only 30 to 40 percent of embryos ever create a successful pregnancy. The problem is not with the sperm meeting the egg, or with the two of them merging to create a zygote—the one-celled entity that combines the genes of both mother and father. Apparently, fertilization is the easy part—it's the next step in which a multitude of variables can intervene. Because of genetically defective zygotes, defective eggs or sperm, or a host of other reasons, most embryos simply pass through the uterus without attaching to its wall, where they must develop the all-important placenta.[3] This is the critical step that establishes a pregnancy. The placenta is a special structure that projects from the wall of the uterus and provides the umbilical cord, nurturing blood vessels, and the amniotic fluid that a baby will develop in. There is no doubt that without implantation and the formation of a placenta, the embryo will never form the vital relationship to the mother's body that is necessary for its development beyond a tiny clump of cells.

In their acclaimed book *The Facts of Life: Science and the Abortion Controversy*, biologist Harold Morowitz and physicist James Trefil note that in the natural process of reproduction:

> Slightly fewer than a third of all conceptions lead to a fetus that has a chance of developing. In other words, if you were to choose a zygote at random and follow it though the first week of development, the chances are less than one in three that it would still be there at full term, even though there has been no human intervention. . . . It is simply not true that most zygotes, if undisturbed, will produce a live human being. The probability that a conception will result in a live birth is actually quite low.[4]

In most cases of natural conception, in other words, the embryo will simply pass through the woman's body, and she will never know it existed.

In the clinic, however, shouldn't we be able to gain better control of the process and avoid the further creation of excess embryos? Many IVF clinics are working toward the goal of improving the ratio of embryos created to embryos that result in a pregnancy. But no matter how much IVF techniques are refined, they will always be faced with the fact that most embryos are not viable. A government-sponsored report to Congress issued in May 2005 estimated that about 60 percent of the embryos created in the IVF process are judged to be incapable of leading to a live birth for one reason or another.[5] These embryos are not considered suitable for transfer into the uterus, but a decision must still be made about their dispensation. Because healthy embryos—those healthy enough to survive the rigors of natural selection—are so difficult to create, the fact that clinics create embryos at all means that it's unavoidable that many of them will not be suitable for reproduction. In other words, excess embryos are all but inevitable. And this is just as true in nature as it is in the clinic. It is because of this high failure rate of embryo implantation that IVF clinics transfer three to four embryos, on average, into a woman's uterus at one time. Even then, more often than not, no pregnancy results from the transfer.

Some critics of IVF go so far as to say that clinics should produce only the exact number of embryos that are needed to produce the number of children a couple desires. But again, this is impossible because there is not a one-to-one ratio of embryos created to embryos establishing pregnancies. It has been suggested that clinics "implant" one embryo at a time, but this represents a misunderstanding of how reproduction—whether natural or assisted—works. Doctors cannot *implant* embryos—they can only *transfer* embryos to the womb and hope that they *become* implanted. Like it or not, there's no getting around the fact that nature imposes its own "survival of the fittest" law on the embryo.

Another limiting factor that prevents clinics from creating just one embryo at a time is the extremely high cost of IVF, which is rarely covered by insurance. Couples pay about $10,000 per IVF cycle (a cycle covers the period when a woman receives drugs to induce hyper-ovulation, followed by the retrieval of eggs and attempts to fertilize and transfer them). Almost no one meets their goals after only one IVF cycle. It can take three or four IVF cycles, transferring three of four embryos each time, before a pregnancy results, while others can go through that many cycles without ever becoming pregnant. If clinics transferred only *one* embryo per cycle, the cost to produce even one child could be astronomical. It would also entail prolonged risk to the mother, who must take fertility drugs to induce ovulation. These drugs are not without their risks, and can have side effects that include abdominal pain and swelling, nausea, dizziness, and headaches. In rare cases, hyperstimulated ovaries can even cause death.

Critics of IVF who demand that clinics either meet a one-to-one ratio or stop doing IVF are asking them to do something that even nature cannot do and that is probably inadvisable as well. After all, one of the main reasons that embryos often fail to implant in the uterus is that they have chromosomal abnormalities. Even if IVF had a "perfect" success rate and induced pregnancies every time an embryo was transferred, there would still be embryos with chromosomal abnormalities, which would not be suitable for uterine transfer. Most parents would not be willing to transfer an embryo that would almost certainly fail to implant, or that would result in a very sick child, so these embryos would simply be added to the total of frozen excess embryos. Until the science of assisted reproduction becomes far more advanced than it is today, and even exceeds natural success rates, there is no getting around the fact that there will be embryos in excess of the number needed for reproductive purposes. And considering that about 10 percent of U.S. couples experience infertility, it is not likely that the demand for assisted reproduction is going to go away any time soon.

So what are the fates of excess embryos? Are they "owned" by anyone? Do they have rights? And if so, who determines what happens to them? Not surprisingly, the egg and sperm donors own any embryos left over from IVF cycles, and they alone can decide what to do with them. Because of the heated politics surrounding the issue, the subject of what to do with all those frozen embryos has become confused by rhetoric that is completely unanchored to reality. One of the chief obfuscations is statements (made even by the president of the United States) that imply that anyone other than the genetic donors have the legal right to decide on the dispensation of embryos. But those who enter into the IVF process sign legally binding agreements that spell out the possible outcomes of the process, and confirm that the "ownership" of the embryos rests with the egg and sperm donors.

President Bush has announced publicly that he believes excess embryos should be "put up for adoption," but even *he* doesn't have the right to dictate a couple's decision. For couples to lose the right to decide what happens with their excess embryos would take an act of law that would nullify the agreements that these couples have entered into with the clinics providing IVF services. Until that happens, the couples (or in some cases, the single women seeking to become parents) will be the decision-makers. This is why, in a February 12, 2006, episode of *60 Minutes*, the bioethicist Art Caplan told Leslie Stahl that the president's position is both "hypocritical and deceptive." As mentioned earlier, one reason the president's position is seen as hypocritical is that even though he claims to believe that destroying an embryo is taking a human life, he is not willing to challenge the practice of fertility clinics disposing of embryos. And his policy of promoting "embryo adoption" is deceptive because there is no realistic possibility of this happening on a large enough scale to solve the "embryo problem."

Under the Bush administration, the U.S. government has awarded approximately $1,000,000 to embryo-adoption "awareness programs," such as the Nightlight Christian Adoption Agency's Snowflakes Program.[6] These organizations have perhaps made more people aware of the option of "adopting" frozen embryos. But unless Mr. Bush plans to send federal troops to every IVF clinic in the country to seize control of the embryos in their freezers, his statements can be nothing more than political pandering to his conservative base. Even assuming that we established an "embryo adoption mandate" in which the only legal disposition of the embryos was adoption, there's a huge problem—the number of willing participants in such a program is miniscule.

There is currently no impediment to couples donating their left-over embryos to others for reproductive purposes—this is one of the options presented to them as a possible use of any excess embryos when they sign an agreement with their fertility clinic. But very few couples are willing to do so. Most people simply don't like the idea of having other people raise their genetic children. As for couples wishing to "adopt" frozen embryos, since 1980, less than 100 couples (or mothers) have volunteered to become surrogate parents of excess IVF embryos. The reason is that couples who are able to be gestational parents naturally prefer to have their own genetic children, and those who are not able to become genetic parents can adopt live children without going through the risk and trouble of pregnancy. It might sound good for the president and for conservative bioethicists to say that they're in favor of putting all the embryos up for adoption, but the reality is that very few of these embryos will ever be adopted.

On the above-mentioned *60 Minutes* episode, Dr. Robert George, a member of the President's Council on Bioethics, said he doesn't think couples should have the right to decide what happens to their excess embryos. Without explicitly saying who he thinks should have the right to decide, the implication is that perhaps the government should have jurisdiction over these embryos. What this policy really means, in effect, is that the vast majority of the embryos will simply sit frozen in their liquid nitrogen tanks until they either deteriorate past the point of viability or some other circumstance results in their destruction. But at the rate of adoption so far, a frozen embryo has, at most, about one chance in 4,000 of being adopted.

So if we take the unrealistic option of adoption off the table, we're left with the following four options for the frozen embryo: They could be used by the couple for another attempt at pregnancy; they could be disposed of; they could remain frozen indefinitely; or they could be used in medical research, including embryonic stem cell research. If the couple decides to use their remaining embryos to have more children, the problem is solved. However, this outcome is not likely because most couples who go through IVF want no more than two children. Some of the frozen embryos might end up being used for reproduction, but the possibility that *most* of the frozen embryos will be used this way is nonexistent because many of the couples have already completed their goals for family-building.

Another option is to discard the embryos as medical waste. This happens on a daily basis in clinics across the country. Embryos are simply taken out of their tanks and allowed to thaw. Many of them will not survive the thawing process, but the cells in about half of them will survive for a few days in the test tubes. Within a day or two, the cells will stop dividing, and they will be placed in biological waste containers and discarded. Stem cell scientists and patient advocates bemoan this situation, which benefits no one. Although there's a chance that some of the embryos were frozen because they were not considered optimal for uterine transfer due to the quality of the egg or because the cells did not divide readily, they can still give rise to normal cells that would be useful in research. Again, pro-stem cell research advocates point out the irony in the fact that the disposal of embryos is uncontested by right-to-life groups, while the effort to use them in potentially life-saving research is being vigorously fought.

Another possible scenario is to do nothing, and to allow the embryos to remain indefinitely in a state of frozen suspended animation. In fact, many embryos have been frozen for over a decade, and maintaining them in their cryogenic state costs couples about $2,000 a year. In some cases, people simply stop paying for the embryo preservation, or move away, failing to notify the clinic of what they would like done with the embryos they own. Some IVF clinics spell out in their agreements that they will maintain frozen embryos for only a limited time (usually five years), after which the couple must decide what can be done with them. If the donors can't be located, then after a period of time (which has been established in their contracts) the clinics have a right to discard them. This raises yet another question about the Bush policy of regarding every embryo as a human life—if couples are no longer willing or able to pay the $2,000 preservation fee, should the government take over the payment in order to prevent their "death" or destruction?

Some people feel that keeping embryos cryogenically preserved is a more morally acceptable option than either discarding them or using them in research because it doesn't entail willfully destroying them. Yet in the long run, this lack of a decision becomes a decision in itself. In both the "discard them" scenario and the "keep them frozen indefinitely" scenario, the bottom line is that the embryos are never going to develop into a baby. Pro-stem cell research advocates argue that in this scenario, again, no potential life is lost because the embryo will never be transferred to a womb, yet a precious medical resource is being denied to patients that might benefit from them.

The fourth option is that embryos can be donated for use in biomedical research, for the derivation of embryonic stem cells, and perhaps even stem cell treatments. If the couple decides on this option, the fertility clinic may supply the embryos to a university or other research organization seeking embryos for research. Because embryonic stem cell research has barely begun in the United States, there are so far only a few of these relationships in place.

Louis Guenin, a moral philosopher at Harvard, argues that we must weigh the fact that these embryos have no chance of ever developing into a person with the reality that they could be of service in the amelioration of widespread disease and suffering. He notes that "It is virtuous to eliminate suffering in *actual* lives when we may do so at no cost to *potential* lives."[7] There are no potential lives because the *decision not to transfer* the embryos into a womb is what determines their potential, not the blind hand of "fate" or some other contingency that is beyond our control. The fact that some egg and sperm donors do not want these embryos used for reproductive purposes closes off the possibility that they will ever become people. That is the reality we should act upon, not some hypothetical event that will never occur.

At the moment there is no legal question about who has the right to decide the fate of individual embryos. That decision is in the hands of the egg and sperm donors who initiated their creation. The question that is perhaps relevant in the current political situation is: Should the government have the right to impose restrictions on their decision? This is something that we have every reason to guard against. The dispensation of one's possible genetic offspring is one of the most intimate decisions that anyone could ever face. Allowing anyone other than the genetic parents to decide what can be done with IVF embryos would set a very dangerous precedent in the arena of reproductive free choice. This is why many people become nervous when the president and his advisors start to imply that the government needs to step in and take control out of the hands of the parents of embryos.

Another question altogether is how each of us feels about perhaps receiving treatments that were derived in some way from research on blastocysts. Embryonic stem cell research is likely to intimately affect all of us, whether we're aware of it or not. We should ask ourselves, "Am I comfortable receiving medical treatments at some point in my life that were derived from a lab-created embryo?" Some people are not, and they should have the choice of saying "no" to stem cell transplants or any treatments that involved the destruction of an embryo. But should those who object to embryonic stem cell research be able to impose their choices on others? Many conservative politicians and members of the right-to-life movement say they should. One analogy to this situation would be if Jehovah's Witnesses, who call upon their members to reject blood transfusions, could decide that no one could receive a blood transfusion. Most people would say this is unfair, yet it's an accurate analogy to the current federal policy on embryonic stem cell research.

Because science doesn't have a ready answer to the metaphysical question of whether a blastocyst should be regarded as a human life, we look to our religious and philosophical traditions for guidance. The issue of whether or not embryonic stem cell research is admissible is, for most people, strongly influenced by their religious beliefs about the status of an early-stage embryo. There are many different religious opinions on the subject, and to complicate matters, most religious traditions were forged long before the modern-day phenomena of IVF and stem cell research. But to claim, as some religious extremists do, that one cannot be religious and be in favor of stem cell research is simply not true.

Everyone has heard of the absolutist view that human life begins at the moment of conception. This position is well known because its adherents are already highly vocal through their participation in the right-to-life movement. But to assume that the "life begins at conception" argument is the *only* religious view would be absolutely incorrect. There is already a wide range of opinions on the part of religions, of denominations within religious, and among various religious thinkers about when life begins, and about the permissibility of stem cell research.

The first major school of thought posits that human life begins at the very moment when the sperm meets the egg (or assuming the act of cloning, when the somatic cell is fused with the egg). This is the absolutist view, and it is held by the Catholic Church and by many modern evangelical churches (to be more scientifically accurate, the "moment of conception" spoken of here should really be referred to as the moment of fertilization; in the view of some, conception cannot really be said to have happened until a pregnancy is established in the womb). The other school of thought, which includes the views of many religions, holds that human life begins at the time of *ensoulment*, or when the developing fetus receives a soul (or for the nonreligious, a mind). Most people who are proponents of the ensoulment viewpoint believe that the developing fetus receives a soul or mind at some time during development, usually measured in gestational weeks or months.

The biologist and bioethicist Jane Maienschein calls the two major viewpoints the "preformist" view (which assumes that full personhood exists at the moment of fertilization) and the "epigenesist" view (the view that personhood emerges gradually and only comes into being at some later stage of development). She has argued that "The preformist interpretation, favored by many conservative religious groups today, lends itself most readily to a strictly genetic determinist view."[8] This is one of the most striking points that is lost in the heat of the debate—the fact that the absolutist view of the embryo is also the most materialistic viewpoint. This view places its determination of personhood solely on the physical presence of DNA.

As for those who believe that human life emerges gradually, there is no agreement on when a human comes into being, and we may never come to an agreement about this. As mentioned above, the Catholic Church, prior to the nineteenth century, adhered to the Augustinian belief that a soul emerges at 40 days of gestation (for males, and 80 days for females!). But again, when we look to religion for answers, it's important to remember that phenomena such as in-vitro embryos, stem cell research, and the possibility of therapeutic cloning didn't exist when the vast majority of religious scriptures were written. Many religious organizations, including Christian Protestant churches, are still grappling with the formulation of a policy concerning stem cell research and other new biomedical developments based on their spiritual and theological traditions. The last few years have seen an explosion of new writings from theologians of all traditions on these subjects, some of which may eventually be formalized as church policies. But many of the more liberal religious organizations simply don't issue canonical positions on these subjects—they may always leave these judgments to the personal conscience of the believer.

The view that embryonic stem cell research is universally opposed by those with strong religious beliefs is a major misconception. Millions of deeply religious Americans feel that the potential for the alleviation of human suffering inherent in embryonic stem cell research is perfectly consistent with their values. Several Protestant churches today have formulated positions that are supportive of embryonic stem cell research. For these churches, there is growing disquiet over the religious debate being more or less dominated by the Catholic and more extreme evangelical views. In 2006 in Maryland, where a bill to provide state funding for embryonic stem cell research was being considered, some pro-research churches were starting to make their voices heard in testimonies at the state legislature. Reverend Peter Nord of the Presbytery of Baltimore told the *Baltimore Sun*, "What troubles me is that the current Catholic understanding regarding the beginning of life seems to somehow have become the gold standard by which everything is judged. That is one perspective, but there are others. Most of us support the use of embryos that would otherwise be discarded."[9]

The Presbyterian view expressed by Reverend Nord is shared by the United Methodist Church, the United Church of Christ, the Unitarian Universalist Association, and the Episcopal Church.[10] Many Christians have pointed to Jesus's mission of healing the sick as some of the most vivid passages in the New Testament, and interpret this as a strong message that Christians should rightly concern themselves with the alleviation of suffering in the sick. They emphasize the many passages that deal with healing in the New Testament, such as the words in Matthew 4:23: "And Jesus went about all Galilee, teaching in their synagogues, and preaching the gospel of the kingdom, and healing all manner of sickness and all manner of disease among the people."

In addition, mainstream Protestants are more likely to locate the human "self" in the immortal soul or spirit rather than in the body, which is merely physical and transient. In other words, the nucleus of the self is in the soul. Some theologians are working to reconcile such religious traditions with the recent discoveries of science and medicine. The Lutheran theologian Alan Padgett has brought the logic that we apply to the end of life to bear on the problem of the beginning of life: "At the end of life, there is general agreement that a human person needs, among many other things, some brain activity. When brain activity ends, the human person's life is over—at least in this life. Applying this to the start of life, at least some brain or neural activity would seem to be necessary for an embryo to be a person." He goes on to say, "I prefer to think of the zygote and the proto-embryo as the *seed of a human body* . . . but this does not mean that research on zygotes is automatically immoral. Such research may be justified if done for the right reasons and in the right way."[11]

In the Old Testament, Adam receives a "living soul" when God "breathed into his nostrils the breath of life" (Genesis 2:7). This led to a predominant belief in Judaism, and in many mainstream Christian sects as well, that human life begins with a baby's first breaths. The Jewish faith, in general, is supportive of the use of embryos in research aimed at curing the sick. While there is some difference of opinion about whether or not therapeutic cloning is acceptable, the United Synagogue of Conservative Judaism and several other Jewish denominations support it as long as it is done for the purpose of healing and not for human enhancement (that is, to produce desirable physical or mental traits).[12]

There is little agreement among the world's major religions about when a human life begins, yet some attitudes toward embryonic stem cell research are quite surprising. The Muslim faith, for example, holds that human life begins at 120 days, or four months, after conception. The Koran (39:6) states: "He creates you stage by stage in your mothers' wombs in a threefold darkness." Research with embryos, especially for the purpose of healing the sick, is acceptable. However, Islamic law prohibits adoption and surrogate parenthood, so the possibility of adopting embryos for reproductive purposes is not on the table. Muslim scholars, hold differing opinions about the deliberate creation of embryos through therapeutic cloning for the purpose of research or for medical treatments.[13]

Hindus believe that life begins at conception, but the destruction of an embryo can be measured against a greater good, such as healing those who are suffering. In spite of the religious doctrines of Hinduism, India has an active stem cell research community that is supported by its government.[14] The Buddhist position regarding embryonic stem cell research is fascinatingly nuanced. Early Buddhist scriptures claim that life begins at conception, but some of the most liberal attitudes toward embryonic stem cell research, including therapeutic cloning, have arisen in the Far Eastern cultures of Sri Lanka, Singapore, Thailand, Taiwan, Japan, and South Korea. Although abortion is illegal in South Korea, a majority of the population strongly supports stem cell research, including therapeutic cloning. (It is just as surprising, in my mind, that the Catholic Church is opposed to both abortion and embryonic stem cell research, but not to fetal tissue research, which involves research on aborted fetuses.)

Part of the reason that Asian Buddhists accept stem cell research more readily than some others (including American Buddhists) is that they do not hold to the belief (as some Westerners do) that man-made improvements in matters of health are a direct affront to God. Buddhist ethics are not intended as laws, but as guidelines to which human beings must bring their own rational interpretations when deciding what is right and what is wrong. One is expected to sort through life's moral ambiguities and try to decide the best outcomes for the greatest number of human beings, and the relief of human suffering is seen as an overarching good.[15]

When I meet with people who believe that life begins at fertilization, I am always struck by the fact that this is the most materialistic view someone can take. This is the view that the presence of 46 human chromosomes in a cell, regardless of anything else, makes the embryo equivalent to any man, woman, or child. Official Catholic dogma has abandoned the question of whether an embryo has a human soul, mind, or anything more than human DNA. In 2004, Father Tadeusz Pacholczyk, one of the Church's most vigorous spokesmen on the subject of stem cell research, told the St. Louis Post-Dispatch, speaking of embryos, "We don't care if there's a soul or not, we care if its being is human."[16] In an embryo, we know there is no mind because there is no physical mechanism to support a mind. If one takes away the possibility of a soul, all that is left is cells. In 2005, a spokesman for the Church's Spanish Episcopal Conference, Father Juan Antonio Martinez Camino told the international press, "Where there is a live human body, even if only for a day, it's a person."[17] Here a clump of cells is being equated with an entire body as well as personhood. We can only conclude that it is the presence of human DNA that alone defines the early-stage embryo as a human being. This to me is another way of saying that your DNA is "you." But this is a deeply reductionist way of viewing things. It implies that none of us could exist independently of our DNA.

It is in clear contrast to the spiritual view of human life, which regards the body as the temporary "house of the soul" but not the soul itself or even the essence of the person. After all, when we die, we will leave our DNA behind, as part of the "dust" that returns to dust.

The assumption that the presence of human DNA is the criteria for personhood raises philosophical questions that can only lead to absurdity. As Ian Wilmut has said, "People are not genes. They are so much more than that." Indeed, every cell of our bodies contains a complete set of chromosomes, and if inserted into an egg cell and implanted into a womb, could grow into an entire person. Yet we don't hold mass funerals every time we wash our hands or comb our hair, shedding *thousands* of these cells. Nor do we regard the act of taking a shower as a holocaust in which millions of mini-people are washed down the drain.

Another angle to this argument is that it's the genetic *uniqueness* of the blastocyst that makes it equivalent to a full-fledged person. But there are too many biological exceptions to genetic uniqueness to make it a coherent argument for personhood. For one thing, an embryo at the blastocyst stage can still decide to split into twins, triplets, or other multiples, so the issue of uniqueness is still an open question. And then there's the issue of natural chimeras—no, not freakish amalgamations of different animals developed in the labs of mad scientists—normal people who happen to harbor more than one set of genes. One of those genomes might be shared with another person, living or dead. This is related to another recent discovery in embryology—that many more of us started out in the womb as twins than we ever suspected.

In fact, there is a pretty good chance that you and I first shared our mother's womb with either a fraternal or an identical twin. Some recent studies even suggest that up to 50 percent of very early pregnancies start out this way. I mentioned earlier that fewer than half of all embryos manage to implant in the lining of the womb. When implantation *does* take place (perhaps because the lining of the uterus is in a state of maximum receptivity), sometimes more than one embryo implants, resulting in embryonic fraternal twins. This is the reason many women who receive IVF end up with twins. However, in the first few weeks of pregnancy, there is another brutal sifting of possibilities. One twin "takes," and the other one simply dissolves.[18] But this story gets even stranger than that.

Some of the cells of the dissolving twin can be absorbed into the body of the living twin, and stay there for the rest of his or her life. And very frequently, some of the cells of a fetus enter the body of the mother through her bloodstream, then travel throughout her organs, where they can live for decades. Scientists discovered this phenomenon when they observed cells with male chromosomes scattered throughout the bodies of mothers who have at one time been pregnant with sons.[19] There are also cases of individuals whose bodies are "mottled" with cells of a different genetic makeup than that of their other cells. These natural chimeras are one organism with two distinct sets of DNA. Are these people science fiction freaks with Dr. Jekyll—Mr. Hyde identities? Not at all. They are normal people harboring some of the cells of a vanished fraternal twin. The phenomena of potential twinning, of chimerism, and of the technical potential of every individual cell to produce a complete organism negate the position that it is the presence of human DNA (or even genetic uniqueness) that makes us fully human. This absolutist view also runs into problems when regarding the phenomenon of living identical twins and other identical multiples. Are they less human because they are not genetically unique?

In August 2004, the Democratic Senator Dick Durbin of Illinois said of the Bush decision to restrict funding for embryonic stem cell research: "There are so many different avenues of opportunity that have been closed by President Bush's decision. What is interesting to me is that he makes this argument on moral and religious grounds. Yet if you accept that, it's hard to explain why he allowed any stem cell lines to be used in research." Many critics have pointed out that the Bush policy is morally and logically inconsistent. It's hard to get around the fact that an absolutist view of the embryo is incompatible with IVF practices, yet the president allows them. Endowing embryos with full personhood also leads us to absurd extremes. Robert George, the president's bioethics advisor I highlighted earlier, has said that if frozen embryos from IVF clinics are going to be discarded, they should be buried or cremated "the same as any other human remains." Does that mean people should buy burial plots and hold funerals for their excess embryos? And what about all of the embryos that silently "pass away" in nature?

A belief in absolutes formulated in periods of history when our scientific understanding of life was far less advanced than it is today will not help us answer the bioethical problems we are confronted with. We cannot simply ignore the issue of the frozen embryos, because they already exist. It is our responsibility to decide what options can be offered to the donors, and then to decide what can be done with those that are donated for research. Bioethicist Gene Outka has written, "I acknowledge that the present debates on embryonic stem cell research involves a moral space that is, to a degree, unprecedented." He noted that it is hard to see the disposal of frozen embryos, which benefits no one, as a better alternative than using them to find cures for debilitating diseases.[20]

Thinking in terms of absolutes can only lead us into a state of paralysis over this issue and leads us to do nothing to address our real, far-from-theoretical problem—the largely unsatisfactory state of human health. The fate of a frozen embryo which is not destined for uterine transfer—while we all recognize that it deserves our respect—must be weighed against the possibilities for good to come out of its use in research. The enormous mass of human suffering due to diseases and conditions that might be alleviated through the wise use of these embryos is real, and it exists not in theory, but in actuality. It is estimated that 3,000 Americans die each day from cell-based conditions that embryonic stem cell research might end up curing. And what about those who are locked into an ongoing struggle with chronic disease and disability? This suffering, experienced by millions, can't be measured, but surely we can weigh it against the fact that a pre-implantation embryo does not have even the most rudimentary capacity for suffering.

A sense of proportion must also be applied to deciding which moral problems we devote our energy and resources to. In a world that is full of poverty, violence, crime, and injustice, and where a large number of children worldwide are abused, orphaned, or abandoned by their parents, should we be pouring so much of our energy into a fight to "save" theoretical humans that have no realistic possibility of ever existing? Should we not be far more inflamed over the fate of the children of Darfur in Sudan, who have been left as destitute victims of a genocidal war? Or for another comparison, the *Atlanta Journal-Constitution*'s Cynthia Tucker has written about the current policy on stem cell research, "I certainly don't understand a 21st century superpower that devotes billions to building smart bombs to destroy life efficiently but refuses to fund the research that could save or enhance the lives of millions of its citizens."[21]

To equate a blastocyst with a living child represents a fundamental breakdown in common sense. While some people philosophically equate the potential (no matter how infinitesimal) with the actual, to start applying this concept to everyday reality would be a formula for chaos. We don't live in a world where all possibilities are occurring at once—we live in a world of past, present, and future. Our sense of proportion tells us that there is a huge existential difference between what is potential and what is actual, especially in the area of what constitutes a person. Given what biological science knows now, one could easily argue that any human cell is a potential person, but for sanity's sake, we don't regard every cell of our bodies as a person.

As for the potential of an in-vitro embryo, including the cloned embryo, to become a person, we have no choice but to accept the fact that its fate cannot be divorced from the *human decision* about what to do, or what not to do, with it. Even the decision to do nothing would irreversibly seal the fate of the frozen embryo. Keeping it frozen forever is only deceptively more palatable than discarding it, because it consigns the embryo to an eternity of suspended animation that, for all practical purposes, is indistinguishable from death. And it would exclude the embryo from any contribution to the efforts to cure disease and end real suffering.

For people like Robert George and President Bush, there is an all-or-nothing conviction that every embryo must have every imaginable chance to develop—a chance that greatly exceeds its chances in nature and even entails going to great artificial lengths to realize. But our sense of proportion leads us to ask, wouldn't the financial and medical resources needed for such an unrealistic enterprise be better devoted to living people who are now sick and suffering?

One of the greatest weaknesses in the absolutist argument is that it ignores the unprecedented issues that attend to embryos existing in lab dishes, outside of the human body. The problem is that these embryos are spoken of as if they were pregnancies underway, or embryos already implanted in a uterus. In reality, they're in test tubes or petri dishes—not wombs. The issue of proper dispensation of in-vitro embryos will never be resolved without acknowledging their location, which has a deciding influence over their condition and potential. Their potential to become persons is not an absolute, it is *conditional*—on whether they are transferred to a living womb, and on whether they become implanted there and make the entire journey to birth. Our decisions should also be based on an honest assessment of their possible fates. While it sounds good to say "put them up for adoption," this position is meaningless unless we have a plan for how to execute it. In reality, this position is no different in outcome than the decision to do nothing—that is, to keep them in frozen limbo until a power outage or some other circumstance makes the real decision for us. It is in my view a position of avoiding a real decision. And we should also keep in mind that the Catholic Church, which takes one of the strictest views against embryo research, is also opposed to surrogate motherhood and assisted reproduction, so the alleged feel-good option of "let them be adopted" will not even satisfy all of the opponents of research.[22]

Surely we can summon the common sense, courage, and collective wisdom to decide what is best for the common good. There is an excellent thought experiment that has been shared widely among thinkers on this subject, and it is worth repeating here: If we discovered that an in-vitro fertilization clinic was on fire, who would we save first? The doctors, the nurses, and the patients, or the embryos in the freezer?

chapter eight

hypocrisy and health care

Government restrictions [on embryonic stem cell research have] severely hampered the ability of researchers to pursue the best science and discouraged many bright young investigators from entering the field.[1]

—*David A. Shaywitz, Harvard Stem Cell Institute*

In our new century's most closely watched race, the United States, long the global life sciences hegemony, is indeed falling uncharacteristically behind.[2]

—*Aaron Levine, Princeton University*

Danny Heumann was 18 years old in August 1985, when he climbed into a car with three other teenagers. All of them had just spent their summer vacation as camp counselors at Camp Baco in upstate New York's Adirondack Mountains. Danny had been a camper there for seven summers, and this year he had been promoted to head tennis counselor. While winding up the summer right after high school graduation, the bubbly, energetic New Yorker was full of anticipation for the upcoming fall. He had only two weeks to go before he started his freshman year at Syracuse University.

But before going home to prepare for college, Danny and his friends decided that they wanted one "last blast" to enjoy the summer, before they all went their separate ways to start their freshman years at colleges throughout the country. They were drawn to the beckoning lights of Montreal, where they wanted to spend a few days of freedom in the glamorous city.

The four boys never made it to Montreal. On the evening that they were to leave for the trip, one of the counselor's friends called to say he had gotten lost while driving to meet them, and asked them to wait for him. The friends grew more and more frustrated as the night wore on. At around 11:30 P.M., while they were sitting on a bench near the entrance to the camp, a car screeched by, and one of Danny's friends shouted, "I think the car that just drove by is my friend Jeff!" The four counselors jumped into the car and took off after the car. They were on a steep, winding mountain road, and racing to catch up with the other car. In the blink of an eye, the driver lost control of the vehicle. The car flipped several times and hit several trees before skidding to a stop. The crash was so intense that it instantly killed the driver of Danny's car and one of its passengers. When the demolished car slid to a stop, the driver's body lay on the ground and Danny lay in the wreckage, his body motionless and in excruciating pain. Despite being critically injured, he never lost consciousness. He could tell that his body was swelling, and he had the agonizing sensation of being pricked by "thousands and thousands of needles in my back." When the paramedics and firemen came, they had to cut the roof of the car open with metal cutters so they could lower a back board into the car using ropes and pulleys. Once they finally lifted Danny out of the car, they loaded him into an ambulance and started off on the 45-minute drive to the nearest hospital, which was in Glen Falls, New York. The whole time Danny was thinking, "This can't be happening. It's just a bad dream. I'm going to wake up, and everything is going to be fine." But the terrible reality was that two boys were dead and Danny was seriously injured while, miraculously, the other boy had walked away from the wreck with barely a scratch. There had been no alcohol or drugs involved; just youthful impatience and a winding mountain road.

When he arrived at the emergency room, Danny was taken immediately for a CAT scan of his back and neck. Soon afterward, the doctor on duty came and delivered the devastating news. He told the 18-year-old that his back was badly broken at the T6 vertebra, and that there was no hope that he would ever, in this lifetime, recover the use of his body from the waist down.

"I just couldn't believe it," Danny says now, 20 years later. "I thought when I left the hospital in Glens Falls and returned to New York City to have my surgery to repair the broken vertebrae and start my rehab, paralysis would be a distant memory—that I'd only be in a wheelchair temporarily. But to add insult to injury, the doctor that night told my parents that I would never father my own biological child."

He spent the next year recovering from the catastrophe, but the following year, he was determined to enter Syracuse University as planned. In his freshman year, while he was taking first-year classes at Syracuse, he was also learning how to live life as a paraplegic. A few years later, for his graduation ceremony, Danny wanted desperately to be able to walk across the stage, to receive his degree standing up just like everybody else. So he spent a year in agonizing training, to build up the necessary endurance to move himself across the stage using leg braces and crutches. It was one of the hardest things he had ever done, but he made the distance without his wheelchair. This was a rare feat for Danny, which depended on the strength of his upper body; he has yet to regain his ability to walk.

Today, Heumann is married and lives in Ann Arbor, Michigan. He devotes much of his time to the Daniel Heumann Fund for Spinal Cord Research, which has raised $5 million for cutting-edge research at universities and foundations all over the world. He is also a powerful motivational speaker at corporations, universities, colleges, and conferences. Danny met Bernie Siegel in 2004 in Miami, where both had been asked to speak at a meeting of the Miami Project to Cure Paralysis. Siegel was inspired by Danny's passionate plea for more research to help people like him walk again, and Danny was fascinated by Bernie's incredible feat at the U.N. They connected after the meeting was over, and Siegel invited Heumann to speak at his next major project—a meeting of some of the world's most prominent researchers at the U.N. This time Bernie had invited scientists from around the world to personally address the General Assembly about the promise of therapeutic cloning. To the profound displeasure of the U.S. delegation at the U.N., Siegel had negotiated a sponsorship of the conference by the 53-nation Asian Group of Legal Experts, and the unusual event was held in the Dag Hammarskold auditorium at the U.N. headquarters. Danny went to New York, where he told the U.N.'s General Assembly how critical it is to people like him for legislators to allow nuclear transfer and embryonic stem cell research to go forward.

From the first time he met him, Bernie recognized that Danny, in spite of his paralysis, is a powerhouse of energy and determination for the cause of stem cell research. "I realized his potential because he's such a powerful speaker," Bernie says. "I told him early on that he's going to be the leader to overturn a bad policy—Michigan is going to be the first state to overturn its bad laws." Since then, Danny has indeed created a large network in his state of politicians, business leaders, philanthropists, and patients who are working aggressively to turn the tide on anti-research legislation. He credits much of his drive in the endeavor to the inspiration he derived from Bernie's belief in him.

I had a chance to talk with Danny at length one day at the University of Michigan (UM) Ann Arbor campus, where I had been invited to speak at a biotechnology conference. It was a Saturday and Danny had graciously agreed to meet me at the university, a place that he was quite familiar with.

The conference I was speaking at was hosted by the University of Michigan's Life Sciences Institute, yet it was evident that one of the most promising fields in biotechnology was not altogether welcome there. In my hotel room the night before the conference, I perused a copy of the next day's schedule, noticing that in an all-day lineup of speakers on biotechnology, I was the only one addressing the topic of stem cell research. I was surprised to see that Michigan Congressman Vernon Ehlers was giving a keynote address. Ehlers had spoken out against embryonic stem cell research, especially therapeutic cloning. Usually, politicians give keynote addresses at conferences like this to act as promoters for the state's biotechnology sector—to more or less get the word out that their state or district welcomes the innovation, prestige, and investment dollars that biotech companies bring. In a state where much of the research is illegal, I wasn't sure how the congressman would approach the issue of stem cell research.

It was late fall and the next morning, the weather was so exquisite that I walked the ten blocks from the hotel to the building where the conference was being held, savoring the crisp but still balmy air. The Ann Arbor campus is the very picture-book image of an American college town.

Old, Tudor-style buildings and stone cathedrals with ivy-covered walls mingled with the newer buildings. Generous, sloping lawns were strewn with the leaves of trees that blazed with bright bursts of red, fiery oranges, and translucent yellows. The deep, sonorous ring of church bells hung over the sidewalk cafés and trendy shops that I walked past, where students were gathered in pairs or small groups to enjoy the glorious weather. However, beneath Ann Arbor's collegiate exterior, one of the most critical battles of the stem cell wars was being fought.

The building at the university's Life Sciences Institute, where the conference was being held, was a gleaming new facility that UM obviously wanted to showcase. While this meeting was going on, the institute was also sponsoring a career fair poster section to help students learn more about fields of opportunity in biotechnology. Clearly, the University of Michigan wanted to beef up its biotech profile by investing in this state-of-the-art center and then opening the conference up to the general public. Not only that, UM had recently announced that it was committing $10.5 million to stem cell research. It would seem that things were off to a good start, except for two things. The first problem is that a state law forbids any research that results in the destruction of an embryo. This means that Michigan scientists could not create any new embryonic stem cell lines that might be safe for human transplantation. They could work with the federally approved embryonic stem cell lines, but as discussed previously, scientists have a very limited incentive to do so. They could use the cells only for basic research and teaching purposes; they would not be able to use these lines to develop human therapies.

The other problem impeding UM's ambitions to become a biotech center is a state law forbidding any kind of cloning, nuclear transfer for the derivation of stem cells included. In short, Michigan scientists can only work with the several-year-old, mouse-contaminated embryonic stem cell lines that are far inferior to cell lines that have since been created. Because of Michigan's laws, the ability to create new embryonic stem cell lines through therapeutic cloning or with the use of excess IVF embryos is impossible even in the private sector, with private money. Because none of the federally approved cell lines are suitable for human transplantation, it will be all but impossible to develop human stem cell therapies in the state. Any biotech companies that want to conduct human embryonic stem cell research have every incentive to move to other states where the laws are not so crippling.

Of course, there are many important avenues of research and development in biotechnology, but it's hard to see a future thriving industry in biotech that doesn't incorporate what scientists think is the most promising biomedical research of our time. And for a university to attract the most promising students and research faculty, shutting down the hottest area in research is a serious problem. In fact, at the time of the UM conference, many American universities had already started to lose graduate students and faculty to California and to foreign countries that fully support stem cell research. There was a conflict between UM's ambition to be a biotechnology leader and a realistic assessment of where the promise lies.

I met Danny briefly just before Congressman Ehlers' keynote address. He wheeled up to me outside the auditorium where the conference was going on and greeted me enthusiastically, wearing a gray fleece jacket and a baseball cap. We spoke for just a minute or so, and then Danny moved to a back-door entrance to the auditorium that was wheelchair-friendly. I went back into the auditorium and introduced myself to Congressman Ehlers, a very distinguished older man who shook my hand warmly. He had the natural grace and friendliness of a seasoned politician. But I couldn't help wondering how he can justify, in his own mind, denying hope to people like Danny Heumann.

Congressman Ehlers was a captivating speaker. He spoke eloquently about the importance of science education—why American students need to be better educated in science, and how our mediocre academic standing vis à vis other countries threatens our economic future. He deplored how few young Americans go into scientific fields and pointed out, humorously, that "In high school you kind of look down on the nerds, but when you get out of high school, you'll either *be* a nerd or you'll work for one." He recognized that in order for the United States to maintain its preeminence in science and technology, we need public policy that promotes partnerships between university scientists doing basic research and biotech companies that have the ability to apply the new technologies to products. He moved on to the problems of ensuring that government policies support scientific research. Here, he said, is where a "scientifically illiterate" public interacts with "politically clueless" scientists.

Of course, most scientists are *not* experts in politics. They are trained for many years in a field that is, or should be, radically different from politics. Scientists are not trained to be politicians, but the more examples I see of stem cell researchers being forced to spend an inordinate amount of time defending their work, the question springs to mind—should they be forced to do this? There are a few scientists who, in addition to being active researchers at the top of their field, are also excellent ambassadors for the cause of stem cell research. Two of them are John Gearhart at Johns Hopkins University and George Daley at Harvard. Both of them not only have incredibly demanding scientific careers—they travel constantly, speaking at meetings, seminars, public gatherings, senate hearings, and congressional briefings, and give interviews to the press on a regular basis. There's no doubt that they have helped immeasurably in the effort to educate the American public and our elected officials about what stem cell research is and how it works. But is it reasonable for us to expect our best scientists to regularly put their research on the back burner while they engage in such an endless whirlwind of promoting the cause of research? Wouldn't their time be better spent doing the research itself?

While Ehlers brought up important points, it seemed to me he was dancing around the issue of stem cell research. He didn't say a single word about it. I was disappointed that he ignored one of the most vital subject areas where science policy is failing us to the greatest extreme. Also overlooked was the critical issue of scientific illiteracy on the part of our politicians, where, in my view, there is no excuse. I can understand that many Americans, whose days are marathons of working, commuting, driving the kids to their lessons, and somehow still getting dinner on the table, aren't going to be brushed up on the latest science impacting our lives. What can't be excused is the politicians whose job it is to be better informed than we are, so that they can make decisions that are in our best interests.

After Ehlers finished speaking, it was my turn to try to cram in an overview of stem cell research. After I finished speaking, Heumann met me outside of the auditorium, and we moved down a wide hallway to find a quiet place to chat. From my vantage point, Danny was silhouetted against a huge window, from which I could see two of the impressive new buildings of UM's Life Sciences Institute. The six-story facilities sat there in pristine condition, their offices still empty. The new complex was clearly a huge investment for the university. All told, the buildings of the Life Sciences Institute offer about 500,000 square feet of space and include a brand new $100 million lab facility. The university's website touts its goal of making the center a mecca for top-notch researchers in molecular and cellular biology, genetics, proteomics (the study of how genes make the proteins that make almost everything else in our bodies), and other forms of cutting-edge biology. Yet I wondered how, given the legislative landscape of the state, UM planned to attract new researchers and investment dollars to fill up those immaculate spaces. I asked Danny what he thought of the university's plans.

"It's really frustrating to me," he said, "because I drive around here and see these beautiful buildings. I see the vitality and the potential of what could be going on here. But we have such restrictive laws on our books here in Michigan. How are we going to be able to recruit the best doctors, the best students? When people graduate from here, a lot of them are going to move to other states, where the laws allow medical progress."

Danny described how, after his fateful meeting with Bernie Siegel and his speech in 2004 for the United Nations, he decided to find out what was happening in his home state. And he was appalled by what he found. "I couldn't believe what I read," he recalled. "I found out that with all this research going on in Ann Arbor, we're the most restrictive state in the country regarding stem cell research."

Since then, he has devoted his energies to trying to affect a change in policies. He contacted a member of the Michigan House of Representatives, Andy Meisner, and told him that he wanted to overturn the laws that forbid embryonic research and therapeutic cloning research. Andy listened. He agreed with Danny, promised his support, and has since been what Danny calls his "champion" in the state legislature. By now Danny was learning from Bernie Siegel and from advocates in other states that if you want to undertake a law-changing initiative, you need to enlist as many partners and allies as possible. He set up a meeting with Meisner and one of UM's attorneys and a university government relations official. They asked the University of Michigan to partner with them by endorsing some pro-research legislation that would overturn the laws criminalizing therapeutic cloning and banning research with embryos. They wanted UM to follow in the footsteps of other universities like Harvard, Stanford, Johns Hopkins, and numerous others that have helped push for reform in their state legislatures. But UM wasn't interested. "Their response to me was, 'We're doing stem cell research here at Michigan, and we have no problems. We're just using Bush's federally approved cell lines.' "

When Danny told me this, I couldn't reconcile it with the university's investment in their new life sciences center. "I just don't get it," I told him. "All the money they've poured into these huge new facilities, and what do they plan to do with them? Just work with the contaminated cell lines? Or maybe focus on adult stem cell research?"

Since his initial contact with officials at UM, Danny replied, the university has started to change its attitude. They have lost one of their best stem cell biologists, Michael Clarke, to Stanford University in California. "Now they've started to see the light, that we've got a real problem here. They're starting to see how these laws really hurt their goals. They've built these huge facilities, and now they need to recruit the best researchers, and it just isn't happening. What we're experiencing in Michigan is a brain drain, with the best scientists leaving and taking their research to other countries or states. This university has a rich tradition of being a research leader in this country, but unfortunately, we're in a situation now where the legislation is preventing the recruitment of scientists."

He went on to describe what *should* be happening. "The key is you build these types of facilities here, you recruit the best scientists, and this is a green light for biotech companies to come here and partner with UM researchers. Then we get research out of these labs which would pay substantial royalties to the university, and biotech companies can take the technology and develop it to where it needs to be to get FDA approval. And finally, the patients who are sitting here (like myself), can benefit from the new therapies."

"How does it make you feel," I asked him, "as a patient—as a person whose life might be radically different—to live in a state where much of the research is illegal? Essentially, your state is saying . . ."

"They're saying to me, to hell with you," Danny replied. I guess there was no diplomatic way to say it, and his anger and frustration were palpable. "It's painful. But they're saying that to the wrong guy, because I'll go out and make speeches and tell my story, and I'll motivate other people to help make a change. When I make a presentation, I give you my heart and my soul. I wear my heart on my sleeve. But at the end of the day, that's how you motivate people. I'm a private citizen who's not a celebrity, who just won't settle for the medical establishment's prognosis that I should spend the rest of my life in a wheelchair."

Not surprisingly, Danny follows the developments in stem cell research closely. He is aware of the experiments in which embryonic stem cells partially reversed spinal cord injury-induced paralysis in rats. I asked him if he thinks that research with embryonic stem cells will ever cure him.

"Maybe," he said. "It's too early to tell. Let's see if this is the magic bullet. And if it is, then bring it on. But we'll never know if we're not allowed to do the research. They [opponents of the research] love to say that with embryonic stem cells, nothing's been done. Well, nothing's been done because of all the restrictive laws. Let's see the promise of embryonic stem cells—let the scientists do the work they need to do to get the job done. And let's see if it's all for real or it's not. But we'll never find out if we don't do the research."

The struggle to enact pro-research legislation in Michigan has been an uphill battle in a House of Representatives that is dominated by Republicans. In 2005 Heumann and Meisner began considering a ballot initiative similar to the one passed in California that added an amendment to the state constitution allowing funding for the research. A ballot initiative would allow Michigan rank-and-file voters (instead of its legislators) to decide on whether or not they want the state to provide funding for embryonic stem cell research. However, getting an item added to the state election ballot is an expensive proposition, involving extensive public education efforts and the gathering of hundreds of thousands of signatures on a petition. Heumann and Meisner started reaching out to local philanthropists to try to raise enough money to do it. Already, they've enlisted support from Michigan's Democratic Congressman Sandy Levin, and Congressman Joe Schwarz, a Republican from Battle Creek who is also a physician.

While Danny and I were talking about the prospects of getting a Michigan ballot initiative off the ground, his cell phone rang and I wasn't surprised when he said he had to go. It was from his wife, Lynn, who had expected him home over an hour ago. And what was so important for him to rush off to so suddenly? It so happens that the day I met with Danny was the day of his daughter Katie's second birthday party. True to form, he refused to accept the emergency room doctor's verdict that he would never father his own child. Two years ago, with the help of in-vitro fertilization, Danny's wife gave birth to a beautiful baby girl, and he wasn't about to miss her birthday party.

Danny's story resonated with me long after he left. It's impossible to talk with him and not realize that, in a couple of heartbeats, with the screech of a tire and the sudden slamming of metal objects, we could all be exactly where he is. We take our lives and our health for granted, but would we be willing to accept a life sentence in a wheelchair, or worse?

As I savored the long walk back to the hotel, I wondered what the odds are that Danny, and the thousands of people like him, will ever be able to enjoy such a simple, joyful experience as a long walk on a crisp fall day. I also reflected on the issues that Congressman Ehlers brought up, wondering how someone could be so aware of the fundamental problem of better educating the public about scientific issues that will intimately affect them, and still be on the side of suppressing a vital branch of medicine.

If America is to have a true democracy, the public must understand much more about recent scientific research than it currently does. This is made more difficult by the fact that some of the most important research happening in labs today is so new that you'd find very little, if anything, about it in existing biology textbooks. We live in such a highly developed research environment that the only way to stay on top of the latest developments is to read the reams of scientific papers published in a staggering array of international scientific journals, to surf scientific websites around the clock, and to read all the popular books on every science subject. And that *still* wouldn't tell you about everything that's going on in the private research sector, where scientists have a commercial disincentive to publish results that they might hope to patent. The two-career couples and busy families of today cannot be expected to spend huge chunks of time perusing *The New England Journal of Medicine*. That's where organizations in the nonprofit sector, such as the Genetics Policy Institute and other educational foundations can fill in the gap—by passing new information on to the public.

Stem cell research is also a perfect example of why it's critical that the public at least understands the basics of the science, because it is something that we will all be voting on, whether we're aware of it or not. The odds are that in the next few years, we could be voting on a state ballot initiative regarding whether or not our state is to fund embryonic stem cell research, or whether therapeutic cloning will remain legal. If we aren't asked to vote on the issue at the state level, the politicians we send to Washington will, without a doubt, be voting on some stem cell research legislation— whether to fund it, promote it, criminalize, or ignore it—and the results of that legislation are bound to have a profound impact on all of our lives. While we usually vote for politicians based on their positions on a variety of issues, we should all remember that if we or someone close to us has dia-betes, cancer, Parkinson's, heart failure, blindness, AIDS, or brain damage from a stroke, that issue is going to eclipse every other issue in our lives. And if we live long enough, that will hold true for each and every one of us.

Danny Heumann's struggle to have embryonic stem cell research con-ducted in his state is only one example of the battles that are going on across the United States. Very little has changed in terms of funding for the research since November 2004, when the citizens of California voted to pass their state's Proposition 71 ballot initiative. The initiative, which passed by a large margin, was meant to establish a creative "work-around" to federal funding restrictions with a state-funded investment in research that far exceeds the contribution of the National Institutes of Health and even rivals the funding in many countries. Through the issuance of state bonds, California allocated an investment of $3 billion over the next ten years for stem cell research, including embryonic stem cell research and therapeutic cloning. As of this writing, four other states (New Jersey, Connecticut, Illinois, and Maryland) have agreed to provide funding for embryonic stem cell research, and activists in several other states are now struggling to establish funding in their own states.

While infusions of cash from the various states are welcome and even necessary developments in the short run, creating a patchwork system of laws in regards to stem cell research over the long term could be very problematic for patients and communities. If no coherent federal policy to support the research is put into place, along with nationwide laws to protect it from anti-research minorities, the result could be a Balkanization of U.S. health care. In a nation where a lack of universal access to health care is already a serious problem, state-by-state laws governing the many treatments and technologies created through stem cells will only exacerbate the inequality.

If inconsistent legislation exists when stem cell therapies become available to patients, there will be more medical "haves" and "have-nots," states where patients can get life-saving treatments, and other states where patients seeking the same treatments would be branded as criminals. Some of the state laws banning or prohibiting the research also brand their own citizens who travel to other states to receive stem cell treatments as criminals. Some states are also considering laws that hold doctors, nurses, and hospitals criminally responsible if they deliver treatments that are outlawed. One of the most infamous of proposed laws was narrowly averted in Texas in 2005. Dubbed the "Granny Goes to Jail Law," this law would put doctors in jail for even discussing stem cell treatments with doctors in another state, and would criminally prosecute relatives who drove their loved ones to medical appointments in which a stem cell treatment is used. And even without state laws to prohibit patients from going to other states to receive treatments, what about those who can't afford to travel to another state for treatments? The patchwork system of funding and legislation would force Americans to accept the idea that life-saving cures will be available to some while denied to others simply because of where they happen to live.

In all the smoke and crossfire of the nationwide battle to allow stem cell research, there is another compelling issue that has received far too little attention: the economics of health care. With the unprecedented aging of the American population, the cost of health care threatens our solvency like never before. We simply can't afford the same disease statistics in baby boomers that characterized their parents' generation.

Widespread chronic illness goes far beyond a medical and scientific problem, and beyond the physical suffering of its victims. If we can put the enormous problem of human suffering aside for a moment and consider the cold, hard economics of the matter, chronic illness now takes a staggering toll on families and on society itself. The state of health care in the United States is enormously complicated by issues of access and affordability. To understand the true impact of such high rates of chronic illness, we have to factor in the reality that millions of Americans have no up-front access to even the limited care that is available to manage their illness. And for practically everyone, receiving long-term care if we become seriously ill or disabled depends on whether we have a loved one who is willing and able to take care of us.

In the United States, most of the care given to the chronically ill is delivered by family members—the parents, wives, husbands, daughters, sons, and other loved ones of the ill and disabled. In 2004, the National Family Caregivers Association referred to America's unpaid and, for the most part, unrecognized family caregivers as an "invisible workforce of 50 million" people.[3] These individuals are the everyday heroes who keep our inadequate, patchwork system of private and public health care going. They provide 80 percent of the home care that is needed on a continual basis, often for years at a time, by the chronically ill. Their services are valued at $257 billion annually. Giant gaps in Medicare, which doesn't cover what it considers "custodial care," and limitations and gaps in private insurance mean that this situation will only get worse as the U.S. population ages and the baby boomers enter their high-risk years for chronic illness.

And what about the personal financial impact of chronic disease and disability? After all, being seriously ill doesn't make the mortgage and the car payments go away. Chronically sick people are subject to the same financial pressures, and often additional ones, that we all face. Unfortunately, many of us are lulled into complacency by the expectation that if medicine doesn't cure us, private insurance or "the system" will take care of us. For a majority of Americans, this simply isn't true. Americans have far less protection from financial devastation due to an illness than most of us realize (or perhaps want to contemplate).

Without a nationalized health care system, having decent health care or sometimes any health care at all is heavily dependent upon one's ability to have a full-time job that offers insurance. The catch-22 is that chronic illness impacts people's ability to work. A job can easily be lost when a person becomes too ill to work, or is fired because being sick causes him to miss too much work. Needless to say, the loss of a job makes it much harder to pay for private insurance, the cost of which is prohibitive for most people. This is free-market capitalism and social Darwinism at its worst. But let's assume for a moment that we have what we think is a good insurance plan through our employer, and we are ill but able-bodied enough to continue working. Or, it is a family member who suffers the catastrophic health problem, so that the person with the insurance can continue working and keeping the health care plan in place. This means that the sick person will get the care that he or she needs, for as long as she needs it, and health insurance will cover it, correct? Unfortunately, in the case of a really expensive or long-term illness, this expectation is completely unrealistic.

We tend to associate unpaid medical debt with poor people, but in truth no one is invulnerable to being financially devastated by a sudden illness or a severe injury brought on by a car accident. A recent study conducted by researchers at Harvard found that more than half of all bankruptcies are caused by illness.[4] An even more recent study conducted by The Commonwealth Fund found that 20 percent of working adults are paying off medical debt.[5] This is partly because prolonged illness often leads to the loss of a job and health insurance, but what many Americans don't realize is that even health insurance may not be protection from financial ruin. Deductibles and copayments can run well into the thousands in the case of a severe illness, and insurance may not pay, or may have a low cap on, essential services like physical therapies after a serious injury or a stroke. The catch is that, if we become seriously injured or sick, most of us *have* to do anything and everything it takes in order to become healthy and functional again, because we can't afford *not* to. With the meter of expensive therapies and treatments constantly running, patients can be forced to rack up astronomical debts in an effort to become functional enough to work again.

To top it all off, there is the health insurance industry's own insurance against bankruptcy, called the "lifetime cap." If you read the fine print in your health care coverage agreement, you will find that your insurance company has probably put a cap on the amount of coverage you can receive in your lifetime. This means that developing a recurring case of cancer that requires successive bouts of expensive treatments or a disease that requires an organ transplant can easily turn a middle-class American into a medical indigent. With surgeries and other treatments easily adding up to six figures, bankruptcy may only be the beginning. Families can lose their homes, their savings, their credit, their children's college educations, and their security in retirement just because one of their members happens to become seriously ill.

So what is insurance good for? Certainly, in the case of acute care for an accident, a life-saving surgery or an erupting health problem, most plans will get you in the door of a hospital and make sure you get the surgery you need. But in the brave new world of limited coverage, you could walk away with a mountain of debt. Just imagine, for a moment, if you or your spouse needs emergency bypass surgery because of an impending heart attack. As more and more Americans are finding, their health insurance makes sure they get the surgery, but through payment caps, denials of coverage for itemized expenses, and refusals to pay for "elective" services (such as physical and speech therapy after a stroke), they can easily end up with $50,000, $100,000, or even $200,000 or more in medical debts. Now ask yourself if, on your current income and with your current level of debt (credit cards, car loans, etc.) you could factor in monthly payments on a $100,000 hospital tab. This is how middle-class families are driven into bankruptcy.

And then there is the plight of the working poor, who are in a terrible bind when it comes to health care. Many low-wage jobs don't offer health insurance, and for those that do, the employee's share of the cost can make it absolutely prohibitive. The rising cost of health insurance not only burdens companies—employees are paying a bigger and bigger chunk of their paychecks for their share of it. It's not unusual for full-time service industry employees making $8 to $10 an hour to be required to pay one-third to one-half of their take-home pay for family insurance coverage. In fact, in 2006, 41 percent of moderate-income, working-age adults have no health insurance, up from 28 percent in 2001.[6] When families have to choose between food, shelter, and health insurance, the availability of insurance is no more than a tragic illusion. This means that many of the working poor simply live with chronic conditions that diminish their quality of life and add to the cycle of poverty by also diminishing their ability to work.

All this adds up to a scenario in which Americans actually pay far more per capita across the board for health care than the citizens of any other industrialized nation. Today, at two *trillion* dollars a year, health care expenditures represent 16 percent of the American economy.[7] Those who have no insurance, currently estimated to be approximately 45 to 50 million Americans, wait until an illness reaches the critical stage before seeking out health care, which they can't afford to pay for.[8] And when they do seek care, they generally show up, very sick, in the emergency rooms of hospitals that receive public funding and are required to treat them. Of course, the cost of treating those who can't pay at even private hospitals is passed on to all of us in the form of rising costs. There is no free health care in the United States, because someone somewhere is paying for it, and probably paying too much. Yet the return we get on this enormous expenditure is disappointing, to say the least.

Perhaps most problematic of all is the fact that, even with the best treatments that current medical science has to offer, more often than not, patients, especially older patients, are not cured. Doctors may be able to treat some of their symptoms, but they might as well be patched together with duct tape and safety pins for all the curing that goes on. To give medicine credit where credit is due, most of us can be kept alive far longer than we would have without it. But even those who have access to the best of care are being kept alive sick, often suffering, and with some level of diminished ability. At the end of our lives, this irony routinely plays out in hospitals to heartbreaking extremes.

These are the tough realities that Americans, with our tendency toward endless optimism and "faith in the system," don't want to face. We have a health care system that promises far more than it delivers, is unevenly distributed (one might even say financially rationed), is enormously costly, and yet leaves one-third of the population ill and grappling with some level of disability. The unprecedented aging of our population under such circumstances has led many public health experts to make dire predictions about the next few decades, including predictions about the collapse of Medicare and Medicaid. And it is beyond dispute that if the incidence of chronic, degenerative diseases remains statistically the same for all age groups, and the total number of Americans over 65 increases to 40 million, as it is expected to do over the next few decades, the burden of disease and disability will create a national fiscal and human disaster.

The rates of illness in the United States are similar to those in many of the world's industrialized nations. Disease and disability are already enormous drains on economies worldwide, but we are at the beginning of the biggest aging boom in history. We need new medical technologies and treatments that don't just manage disease at an exorbitant cost to everyone involved. What we need are healthy people. While there are other bright spots in biomedical research, drugs and devices and other improvements on the horizon, we desperately need to take medicine to the next level. Without the help inherent in the study and use of stem cells, it is hard to see how we might get there. In the next chapter, I'll explore the rocky path that this cutting-edge research has taken in the absence of a major investment from the world's biomedical research giant, the United States.

chapter nine

korea: great expectations

Give me a lever long enough and a fulcrum on which to place it, and I shall move the world.

—Archimedes

The silky gleam of a full moon snaked across the waters of the Sea of Japan as our plane descended. It was a cool October evening in 2005, and I was arriving in Seoul, Korea to attend the opening of the World Stem Cell Hub. The trip from Washington to Seoul is brutally long and exhausting. I left Washington Dulles at 10:00 A.M. on a Monday morning, changing planes in Atlanta. From then on, the Korea Air Lines 747 chased the sun westward across the curvature of the earth for the next 16 hours. This is how one day becomes two before travelers see a sunset. The plane approached Incheon International Airport just before the sun finally sank below the horizon on Tuesday.

After passing through customs, I collected my suitcase, exchanged some American dollars into a large stack of won, and navigated my way

to the airport exit. There I saw a young Korean man holding up a welcome sign that read, "Herold Eve." I greeted him with a surge of gratitude. Looking a little surprised, he said, "I thought you were a man."
We both laughed, understanding that first and last names are in reverse
order in Korea. As we walked, we were quickly joined by an older man in
a suit, who immediately motioned to the younger man, indicating my
suitcase. The young man apologetically took it out of my hand and
pulled it along for me. It was my first taste of the Confucian-style hierarchy of Korean society, which is a far cry from the American way of doing
things.

On the drive to Seoul, in bumper-to-bumper traffic, the older man
drove silently, while my young friend, who spoke broken English, made a
valiant effort to communicate with me. His name was Jae Heon Shin.
I asked him about the research he was working on in the lab of Dr. Woo
Suk Hwang, who was then considered the world's foremost cloning
expert. Dr. Hwang had recently published two landmark papers on
human therapeutic cloning that had made him famous throughout the
world. Fortunately, English is the international language of science, and
we were then on firmer ground conversationally. I was soon fascinated by
the work that this young student was involved in.

Jae Heon was a graduate student at Seoul National University School
of Veterinary Medicine, and he was taking part in some of the most cutting-edge animal research in the world. He worked in Dr. Hwang's lab,
but on a project that is far less famous than the cloning research that had
catapulted the Korean scientist to international fame. Jae Heon was working in research on xenotransplantation, an area that could hold great
promise. Xenotranplantation is a fancy word for cross-species transplant,
such as an animal organ or tissue transplanted into a human. This is by no
means new—pig heart valves have been used to replace failing human
heart valves for years. But several labs throughout the world are now trying to create pigs whose whole organs could be safely transplanted into
humans. There are some major obstacles in this work that need to be overcome before this hope becomes a reality, but animal-to-human transplants could one day save lives by helping to ease the shortage of human
organs.

One of the challenges to using pigs is that normal pig organs are considerably larger than human ones, and the human body cannot accommodate them. Through selective breeding, however, the Korean lab is working on creating a miniature pig, one whose heart, kidneys, and liver would be small enough to transplant into a human. They have been breeding successively smaller animals, which they call mini-pigs, to create more appropriately sized organs. But once the size problem is solved, scientists will still need to tackle another major issue: Pig pathogens can cause lethal infections in human recipients. This danger is all too real. Animals can be infected with retroviruses, or may have old viruses hidden in their genomes that they long ago developed an immunity to. However, once it's introduced into a human body, the virus that was harmless to the animal could be activated and cause a deadly infection.[1] In a worst-case scenario, it could then mutate in the human body and spread not just from animal to person but from person to person, perhaps causing an epidemic that the human immune system would be completely defenseless against. This is exactly what is believed to have happened with the AIDS virus—that it jumped from a primate to a human and then mutated to become transmissible from human to human. As a result, scientists are very careful about the possibility of transmitting animal infections to humans. One of the goals of the Korean researchers is to create what Jae Heon called an "antiseptic pig," one that carries no dangerous germs that could wreak havoc in the human body.[2]

The other challenge to overcome, obviously, is rejection. If humans can reject an organ from another human, it stands to reason that the danger of rejection would be even greater in the case of an animal organ or tissue transplant. In fact, pigs have at least one histoimmune factor (a characteristic that would be recognized as alien by the human body's immune system) that we already know to be problematic. They carry an enzyme that produces a sugar molecule on the surface of their cells that human antibodies will immediately recognize as foreign. Human antibodies will attach to this molecule and go into a red-alert immune reaction that will quickly attack and destroy the organ. Although transplanting pig heart valves into humans has proven to be manageable with the use of immunosuppressant drugs, researchers who have tried transplanting pig organs into other primates found that this reaction was so severe that the organs were destroyed within hours.[3] There's not much point in putting desperately sick people through such a ghastly ordeal, which would almost certainly kill them. But there could be a way around this problem that is not so far down the road. The technique that could be used to overcome this problem is animal cloning.

The Korean lab was taking cells from mature mini-pigs and deactivating, or "knocking out" the gene that produces the enzyme that makes the troublesome sugar molecule. The altered DNA is then fused with a swine oocyte that has had its nucleus removed, and the egg is activated to divide. If the embryo makes it to the blastocyst stage, it is transferred into a sow. If a successful pregnancy results, voila—you have a cloned mini-pig whose cells will not make the dangerous sugar molecule. As exciting as this accomplishment would be, scientists engaged in similar research caution that the sugar molecule may be only one of many issues to overcome in order to make pig organs compatible with humans. No one knows yet when it will be safe to conduct the first pig-to-human transplant trials—if all goes well, it could be in a few years, but if more rejection factors are discovered, it could take several more years to neutralize them. Still, with the use of cloning, scientists now have a huge leg up in the process. Troublesome genes can be knocked out before even an embryo forms, and the new genetic program makes it into every cell of the future animal's body.

Notwithstanding my gut-level response to the idea of raising animals just to harvest their organs, which made me cringe, I wondered why Dr. Hwang, one of the world's foremost stem cell researchers, was even engaged in xenotransplantation research. After all, many scientists think that human stem cells can eventually be used to grow new organs, and through therapeutic cloning, perfectly compatible ones at that. Why bother with all the problems of making cross-species organs compatible, when the world's leading expert in therapeutic cloning could be putting all that energy into growing new organs? Since I planned to interview Dr. Hwang on this trip, I made a mental note to ask him about this. That was the last thought I had just before falling at last into bed at the Seoul Intercontinental Hotel.

In the native language, "Korea" means "Land of Morning Calm." But there's nothing calm about morning, noon, or night in Seoul. In fact, it's one of the most fast-moving, densely packed, and richly textured cities in the world. Modern high-rises and colossal neon signs mix everywhere with the institutions of traditional Korean culture—the rows of small shops and thatched-roofed restaurants receded behind beautiful rock gardens. Frenetic traffic, with its endless streams of cars and death-defying motorcyclists, zooms past old Korean royal palaces with graceful names like the "Palace of Virtuous Longevity." The sidewalks and alleyways teem with fast-walking pedestrians, dressed in the latest fashions and chatting on their cell phones. As winding and complicated as the main streets in Seoul are, they are just the beginning. Behind each street is a second, unbelievably intricate network of secondary alleyways. However, unlike the "dead" alleyways of American cities, which are home to not much more than trash dumpsters, these alleyways are packed with coffee shops, nightclubs, restaurants, and some of the most decadent shopping on earth. Whether you're looking for Gucci bags and Ralph Lauren clothes or just insanely cheap knockoffs, you won't be disappointed—that is, if you have a local resident to show you where to find them.

The Korean love of speed and technology is obvious, yet in most human interactions, Koreans display a charming sense of courtesy and humility. Service is instantaneous, with none of the surly attitude that is becoming a fixture in American life. Those of us who were guests of Seoul National University at the opening of the World Stem Cell Hub were extended every consideration imaginable. Looking back today, I realize what a sublimely proud moment the opening of the hub was, not just for Dr. Hwang and the researchers, but for the university, the hospital, the government, and even ordinary Koreans.

Without witnessing the phenomenon of "Hwang-mania" in Korea firsthand, it's hard for Americans to envision a scientist being lauded about as a national hero, but that is exactly what Woo Suk Hwang was in his homeland in October 2005. Dr. Hwang literally enjoyed rock-star status. He was followed by a mob of reporters, and average, rank-and-file citizens recognized him on the street and asked for his autograph. Once you knew what he looks like—a handsome man with a squarish face who looks younger than his 52 years—you recognized his image everywhere. His picture was on the front pages of newspapers and magazines. He appeared in Korean television musical variety shows sandwiched between famous singers and popular rock bands, in "serious segments" talking about the importance of stem cell research. When I turned on the TV at the Intercontinental Hotel, there he was in a slick, half-hour promotional video showcasing the virtues of his country. Korea Air Lines featured him in their promotional video, and there was even a postage stamp commemorating his research. There's no doubt about it—the entire country was enthralled by Dr. Hwang and exuberantly proud of his accomplishments. He seemed to embody the Koreans' hope for true first-world recognition, for international acclaim, and for a glorious economic future led by Korean science and technology.

At the opening of the World Stem Cell Hub in October 2005, Dr. Hwang was riding high on a wave of achievements that had made him one of the most famous scientists alive. In the previous two years, Hwang and his research team at Seoul National University School of Veterinary Medicine had made history—three times. In 2004, they reported that they had derived human embryonic stem cells through therapeutic cloning—the first research team in the world to do so.[4] As monumental as this feat was, the process of therapeutic cloning was still notoriously inefficient. Hwang and his coauthors in a paper published in the journal *Science* reported using 242 human eggs to produce just one embryonic stem cell line. This was a lot of human eggs, which are well known to be in short supply. Critics of the research in the United States and elsewhere continued to level charges that therapeutic cloning would never be widely applicable because it would require an endless supply of human oocytes, and they suggested that only through the exploitation of poor women, who would be compelled to sell their eggs, would this ever be possible.

However, barely a year after the appearance of the paper in *Science*, Hwang's team, including his American collaborator Dr. Gerald Schatten, reported that they had dramatically increased the efficiency of therapeutic cloning, improving the technique so that it took them only about ten human eggs per one stem cell line created. Not only that, they claimed to have created 11 patient-specific stem cell lines, using donor cells from actual patients with diseases and injuries such as Parkinson's disease, ALS, and spinal cord injury.[5] This was incredible news. Not only did these experiments take the science one giant step closer to providing stem cell treatments that were genetically tailor-made for patients, they also provided disease-specific cell lines for scientists to study in the lab. Diseases in a test tube, so to speak. And if *that* wasn't enough, three months later the team unveiled the world's first cloned dog, a beautiful black and beige Afghan puppy named Snuppy (for Seoul National University puppy).[6]

By this time it was clear that Hwang's team wasn't just lucky—they appeared to have refined existing cloning techniques. Their innovation involved the way that the nucleus is removed from an egg cell. Instead of sucking the nucleus out in a way that also clumsily removes some of the egg's cytoplasm, they first punched a hole in the egg cell's outer membrane with a needle and then gently squeezed the nucleus out. This was less damaging to the egg, and a skin cell from the patient donor could be inserted through the hole in the cellular membrane, which was more precise than just trying to fuse the two cells together. Dr. Hwang joked that his researchers were more successful in their cloning attempts than Western scientists because of the fine motor skills they had acquired through the lifelong use of chopsticks. Whatever led up to it, their success stunned the world.

In 2005, the Korean government rewarded them with millions of research dollars and a brand new, state-of-the-art facility located in Seoul National University Hospital. Hwang's team had reported enormous strides in a few short years, but stem cell research is still not without opposition in Korea. Catholics and evangelical Christians form a minority, but they are strongly opposed to both embryonic research and therapeutic cloning. When I spoke to Dr. Hwang, he described the split in public opinion about his research to be "about the same as in the U.S.—maybe even a little worse." According to public opinion polls conducted in Korea, he said, about two-thirds of Koreans are in favor of the research, while approximately one-third are against it. The difference, however, is that the government's policy is more reflective of the desires of the pro-research contingent. President Moo-Hyun Roh is a strong supporter of the research and, as events were soon to show, the pro-research segment of society is passionate, active, and extremely vocal.

I was one of several guests to attend the opening ceremony of the World Stem Cell Hub (WSCH), the new facility housed within Seoul

National University Hospital. I met up with Bernie Siegel, who was speaking at the attendant symposium, the morning of October 19, and we toured the facility along with a throng of reporters and visiting dignitaries. The hub's opening ceremony was sandwiched between sessions of a biotechnology and bioethics conference, which was heavily attended by businessmen, students, and members of the press. Other English-speaking compatriots who were present that day included Bob Klein, who is president of California Institute for Regenerative Medicine, American bioethicists Laurie Zoloth, Insoo Hyun, and Courtney Campbell; Glyn Stacey, who is director of the UK Stem Cell Bank; and Ian Wilmut, the Scottish scientist who cloned Dolly the sheep.

The opening ceremony itself was held in one of SNU's large auditoriums, an amphitheatre-style room that was lavishly prepared for the event. I was treated to a front-row seat, where the enormous visuals towered over us. Because most of the speeches were in Korean, the English-speakers were outfitted with headphones, which provided running translations. At the front of the auditorium was an enormous, nearly floor-to-ceiling sign, in bright yellow and blue, with the inspiring words "Hope of the World, Dream of Korea," the motto of the WSCH, emblazoned on them. To each side of this monolithic sign were two enormous video screens showing magnified cellular images. A lavish floral arrangement acted as a centerpiece, perhaps to symbolize the blossoming of a new era for Korea.

I was touched by the unadulterated hope and optimism that was so palpable at this event. Koreans regarded the opening of their international center of collaboration as a landmark event for their country. Dr. Hwang's work, and the support for the hub, opened up a whole new chapter in their history, placing South Koreans at the proud center of world events. After a brief introduction by the director of the newly created Seoul Central Stem Cell Bank, Dr. Jung-Gi Im, a video tribute placing Dr. Hwang's discoveries at the apex of modern scientific achievement began. It featured the first flights of the Wright brothers, Alexander Fleming's discovery of penicillin, and Einstein's discovery of the laws of relativity, followed by the Koreans' milestones in therapeutic cloning. It described the World Stem Cell Hub as the "epicenter of world stem cell research." Images of Christopher Reeve and Mohammad Ali were followed by a glorious finale with smiling, happy children, blue skies, and messages of hope.

Immediately following the video, the director of Seoul National University Hospital took to the podium, referring to the WSCH as "a miraculous bedrock contributing to a healthy economy for Korea and health for the world" (direct translations from Korean to English make for some clumsy phrasing, but you get the idea). Dr. Schatten, Dr. Hwang's collaborator from the University of Pittsburgh, spoke next. With his mop of curly gray hair, Schatten looks and sounds like the stereotype of a scientific genius, straight from central casting. He referred to the research climate in Korea as a "biomedical bonfire," and likened Dr. Hwang's journey to a long and lonely climb up a mountain. "But now," he said, "we can glimpse the summit from above the clouds—treatments for some of our most serious disorders." After noting that his mother had died of Alzheimer's disease just one year before, he concluded with, "We now have hope that human suffering may someday be relieved by patient-specific stem cells."

Last up was the country's President Roh himself, a surprisingly young-looking man, wearing a black suit and a pink tie. He confessed that for quite some time, when people talked to him about regenerative medicine, "I didn't really understand. But today, I know it's real." He addressed Dr. Hwang personally, saying, "In the beginning, I didn't give you very much help, but I promise you I will give you more help in the future. My pledge to all of you is more support for basic science." Dr. Hwang later told me that what the president said was true. For years, he had struggled for funding, with very little support from the Korean government. It was only after he published his landmark studies that the government decided to give him the substantial support that allowed for the establishment of the World Stem Cell Hub.

With all the ceremonial rhetoric aside, the WSCH was an ambitious project on a global scale. The scientists who conceived it hoped that through scientific cross-pollination, the hub would serve a critical role in advancing stem cell research (especially therapeutic cloning) worldwide. They recognized that science, more than ever before, is an international effort. This is especially true for stem cell research, considering the radically inconsistent policies of different countries, which have developed varying types of expertise in the field. For example, in 2005 it was thought that Korea led the world in nuclear transfer research, but American scientists were ahead of the Koreans in other techniques, such as directing the stem cells into desired cell types. American scientists had already begun to coax embryonic stem cells into becoming blood cells, neurons and cardiac cells, for example, and U.S. scientists were also on the leading edge of tissue engineering, which holds great promise when combined with stem cells. At any rate, researchers all over the world today recognize that international collaboration and the sharing of knowledge is by far the shortest path to creating cures from the new science.

The hub was designed to act as a disseminator of knowledge among scientists of every country where stem cell research is taking place. Foreign scientists were invited to visit Dr. Hwang's lab, where they could learn Korean cloning techniques, and the hub would also send its specialists to overseas labs to train others. The hub would also be the world's first bank of cloned, disease-specific stem cell lines that would be made available to biomedical researchers everywhere who are trying to untangle diseases. Satellite labs were planned for research-friendly San Francisco and Britain, to provide regional centers for scientists to draw upon.

The United States is a perfect example of why a global network like the one so ambitiously planned by the Koreans could be critical to the development of cures. Even though nuclear transfer is not banned by U.S. federal law, it might as well be. There is no federal funding for nuclear transfer research, and several states even criminalize it. As I discussed earlier, private capital is reluctant to tread where legislation is uncertain, and NIH-funded scientists risk losing all of their funding if they mingle non-government-approved research with the kind that is approved. This makes it incredibly difficult for an American scientist to conduct therapeutic cloning research in the United States. However, if an American scientist travels to a research center in another country, where experts may have mastered the technique, he can learn from them without breaking any laws or violating any Byzantine funding regulations. He no longer has to wait for the political and funding climate to change in his own country before he can take part in the research. At the same time, he can share his expertise with his collaborators, furthering their research. With this system of exchanging knowledge, scientists everywhere can quickly capitalize on the discoveries of others, drastically reducing the amount of time it will take to develop stem cell cures. The hub made such good sense that it's no wonder Gerald Schatten, in a toast he gave at a dinner that night, said, "Woo Suk is like a brilliant spider whose web now encompasses the earth."

But even at the time of its unveiling, it was clear that the idea of the WSCH hadn't been fully fleshed out. When Bernie Siegel and I spoke with Dr. Hwang and other researchers the day after the opening it appeared that there were still uncertainties, and chief among them was the question of patents. One of the most important resources the Koreans were offering the world's scientists is the disease-specific stem cell lines. Having cloned human cells carrying various genetic or degenerative diseases would be incredibly valuable to other scientists. Certainly the Koreans' technique of cloning the stem cells was patentable, but what about the stem cell lines themselves? The SNU researchers, even at the official opening of the hub, seemed to be in the dark about how the patenting issues would be handled by the hub. Since then, there have been conflicting stories dealing with WSCH patents in the Korean and international news.

Dr. Hwang and I finally had a chance to talk at length the night of October the nineteenth in a conference room at the Intercontinental Hotel. The day had been packed with speeches, meetings, dinners, and a reception, and I was amazed that he still had the energy to talk with me about his work.

The whole encounter had a rather formal and official feeling to it. I was told which room to meet him in, then led by several extremely courteous attendants (a row of them, each one bowing in turn) to the room, where I found Dr. Hwang and two young members of his research team sitting at a long conference table. At least part of Dr. Hwang's secret for keeping such a frenetic schedule of researching, writing, traveling, speaking, giving frequent media interviews, and supervising a staff of over 60 people seemed to be that he slept very little, and everything he did, he did quickly. He answered my questions graciously, but when he was finished answering, he would say, "Next question, please."

I asked him whether he thought his cloning techniques would ever be applied to human reproductive cloning. He emphatically told me that any attempt to clone a human baby would not only be unethical, "it would be criminal." The health problems due to genetic abnormalities in cloned animals are heartbreaking—and that's *if* both the mother and the offspring survive to the point of birth. Hwang mentioned the frequent occurrence of a deadly syndrome that is common in animal pregnancies in which the mother is carrying a cloned embryo. This condition causes fetuses to become abnormally large, killing both the mother and the fetus—a consequence so frightening that even the risk makes human cloning attempts unthinkable. But Dr. Hwang had other reasons that convinced him that human reproductive cloning is simply not possible. In fact, only a few weeks earlier, he and Schatten had been in Washington, DC, visiting members of the U.S. Senate, to tell them why they think the specter of human cloning will never materialize—at least not in our life-time.

In April 2003, Dr. Schatten and his research team in Pittsburgh published a paper in *Science* detailing the results of their ongoing attempts to clone humanity's closest relatives—nonhuman primates. Using some of the techniques pioneered in Dr. Hwang's laboratory, they had produced 135 cloned rhesus monkey embryos, which over time were transferred into the wombs of 25 surrogate rhesus mothers. The results were dismal—not one of the embryo transfers resulted in a successful pregnancy. By examining the embryos, the researchers were able to elucidate the reason. It turned out that the nuclei in primate egg cells contain key proteins that are critical for driving embryonic development. Removing the nucleus, which has to happen for cloning to take place, also removed these proteins, arresting embryonic development. The embryos couldn't develop—even in the ideal environment of a womb.[7] The Schatten team also reported that they had strong evidence showing that these proteins exist in the nuclei of human egg cells, meaning that therapeutic cloning to produce a blastocyst is possible, but because the cloned embryos cannot develop beyond a few days, human reproductive cloning is not. At the very least, current science is not any-where close to solving this problem, even if human cloning was the goal.

Dr. Hwang, who was a collaborator with Schatten and his team (three researchers from Dr. Hwang's lab had already been sent to the University of Pittsburgh to teach the American team their cloning techniques), was also convinced that the nuclear protein problem exists in humans as well as in other primates. In fact, he didn't think that the products of human therapeutic cloning should even be called embryos, because they have no potential to become a human life. Instead, he called the cloned embryos "nuclear transfer constructs," useful for the derivation of embryonic stem cells, but with no potential to become a complete organism. If Hwang and Schatten were right, and this could be confirmed by other researchers, it would change the entire debate about human therapeutic cloning. Those, like Senator Brownback, who oppose it would have lost their argument that nuclear transfer creates a human life (even a potential one) and then destroys it. In other words, there would be no question of cloned human embryos having the potential to develop into a baby.

I asked Dr. Hwang about the then-recent report from scientists at Harvard who believed that they had successfully "reprogrammed" adult skin cells by fusing them with embryonic stem cells. The hitch with the reprogrammed cells is that they had two complete sets of DNA, and Dr. Hwang was not optimistic that scientists will find a way to remove an entire set of genes and still end up with a normal cell. "In my opinion, the only real hope for cellular reprogramming is nuclear transfer," he said. I also asked him why his lab was doing research in xenotransplantation—the mini-pig organ donors—when it is widely believed that human stem cells will eventually be used to grow whole organs, and genetically compatible ones at that. His answer surprised me. He didn't think that scientists would be able to grow human organs outside the body from stem cells, as has been widely suggested. "I don't think we will be able to grow organs," he said. "Organs require many cell types, and also the correct structure. Stem cells will only produce small parts of organs. They will never replace organs themselves." Not every scientist would agree with this, but it did explain why the Koreans were still avidly pursuing the possibility of whole organ xenotransplantation. Growing transplantable organs in animals, of course, is a much easier thing to do than growing them in a lab.

At any rate, in spite of his claimed success at cloning human stem cells, Dr. Hwang's animal research was still going strong, and he had no plans to abandon it. He is, after all, a veterinarian and a specialist in animal cloning. "I want to continue to develop cloning technology to supply human organs from donor animals, to preserve endangered species, to make cows that will not develop mad cow disease," he said. Finally, I had an opportunity to ask him about Snuppy, the cloned dog. I had seen photos of Snuppy and he was indeed beautiful, but I couldn't help worrying about his physical health.

"He is very cute and very healthy," Dr. Hwang replied. "I will give you a very special case to meet him tomorrow." I could barely contain my excitement—a meeting with the world's first cloned dog! "But I warn you may fall in love with him!" Seeing the twinkle in Dr. Hwang's eye, I guessed that my cover as an animal lover was blown.

Perhaps what was most significant about our conversation was that Dr. Hwang confirmed that Korea had recently instituted legal regulations to govern embryonic stem cell research, including therapeutic cloning. In January 2005, the Korean National Bioethics Committee's guidelines were established, with strict rules to cover egg donations and the practice of therapeutic cloning.[8] Unlike most countries, Korea had outlawed reproductive cloning and had instituted several regulations concerning nuclear transfer. These rules included a ban on paying women for eggs, and they were backed up by penalties that ranged from the loss of government funding to steep fines and prison time for more serious infractions. In their new regulations, the Bioethics Committee as well as the university's institutional review board can inspect a scientist's records and research techniques at any time to ensure that the rules are being followed. However, these regulations were put into place after the famous breakthroughs that Dr. Hwang reported, and, it could be argued, in *response* to those announcements.

The next morning, Bernie Siegel, a group of German businessmen, and I were treated to a tour of Dr. Hwang's lab. First we were taken to Dr. Hwang's small, cramped, and cluttered office, which exhibited all the signs of an incredibly busy person. Every available surface was covered with stacks of papers, books, scientific journals, awards, mementos, and photos. Dr. Hwang bustled in and out of the office. When he finally summoned us to follow him to the lab, it was clear that there was no time to waste.

On the way to the research lab, we entered a corridor with a wall-mounted lightbox displaying some rather fuzzy black-and-white images.

We had been joined by the German guests, and by Jim Thomson, the secretary of science and technology from the British Embassy in Seoul. Upon closer inspection, I could see that the fuzzy images on the screen were cells, and Dr. Hwang began manipulating the images by pushing buttons on a control panel just below the screen. "These cells are from an eight-year-old patient with spinal cord injury," he said. "These were cloned with no mouse feeder cells." He switched to another image and said, "These are cloned cells from a 14-year-old type 1 diabetes patient." He went through one milky and (to my eyes) indistinct image after another, apparently recognizing each one immediately as though it were the face of a friend. "These cells are from a 46-year-old Lupus patient . . . a 32-year-old with spinal cord injury . . . a 16-year-old male with type 1 diabetes . . ."

Finally one of the businessmen said what everyone else was thinking, which was, "You seem to know all of your cell lines as if they were your children."

"They *are* my children," Dr. Hwang said matter-of-factly. "And please note that there was no difference in proliferation rates based on the age of the cell donor. The cells cloned from the 56-year-old divided just as actively as those cloned from the eight-year-old." Incredible, I thought. It seemed to confirm that even aged cells can be returned to an embryonic state. There wasn't much time for reflection, because Dr. Hwang had already walked briskly away and was about to go through a glass door that clearly led to some kind of transitional space between the hallway and the laboratory.

While a lab such as Dr. Hwang's is not completely sterile, it's important to keep it as free as possible of things like dust, dirt, bacteria, and molds, which can contaminate biological materials, including cells. Before proceeding, we were asked to take off our shoes and put on one of the many pairs of rubber slippers that were lined up against the wall. The slippers, while not sterile, would not introduce as many impurities from the outside world as our shoes would. Once the glass door closed behind us, we were obliged to pull on a sky-blue sterile suit over our clothes. Add to that a sterile cap and a face mask, and the transformation was complete. In small groups of three or four, we then entered a small chamber that resembled a phone booth. Once inside, we were suddenly pelted with powerful blasts of air that were meant to blow every speck of dust off of us.

In the following chamber, we were greeted by seven or eight lab researchers who were sitting around a long table. At first glance, they appeared to be having lunch. The table was littered with containers of varying sizes, and it looked as though each worker had his food spread out in front of him. Two thoughts collided in my head: "Isn't it a little early for lunch?" and "After all of the preparation, they bring *food* into the lab?" Then I realized that sitting on sheets of aluminum foil in front of each worker was a piece of raw meat, and in the scientists' hands were not forks or chopsticks but syringes. "Here they are extracting eggs from cow ovaries, which we get from the slaughterhouse," Dr. Hwang informed us. The ovaries were pale, like chicken breasts, and almost as large, and the scene was a bit unsettling. However, it made sense that the lab would draw upon an industrial source of eggs for its animal experiments.

Next we entered a darkened room where two female researchers were sitting at benches, looking through microscopes. Fortunately for us, each one had a large plasma screen on the wall above her work space that showed what she was doing in greatly magnified form. Egg cells were clearly visible on the screens, with their barely perceptible outer membranes and gray interiors, full of dark and light spots (known as the cell cytoplasm). You could clearly see the dark, round bodies that constituted the egg cells' nuclei. On the screens, the tiny glass pipettes (actually the size of needles) looked huge as the scientists manipulated the slippery eggs. We watched in fascination as Dr. Hwang explained their movements. My eyes were fixed on one of the eggs being held against the end of a pipette while a glass needle gently punctured its membrane and then pressed against it. All of a sudden, a little gray sphere shot out of it—the nucleus of the egg! Before our eyes, an egg cell had been enucleated, its cytoplasm left intact. It was now ready to receive an entire skin cell, or another body cell from a mature animal, for which the entire developmental program would be started all over again. Somewhere in those little gray dots of the cell's cytoplasm was the magic formula that could take any DNA and return it to a primordial state, from which a whole new individual would spring. It was—and still is—an unfathomable mystery.

When we came out of the lab, Bernie and I encountered Dr. Byeong-Chun Lee, a professor at the College of Veterinary Medicine and a person whom I had recently met at a reception in Washington, DC. Dr. Lee, an exuberantly warm and friendly man, had been introduced to me as "Snuppy's father" by Dr. Curie Ahn. Dr. Ahn is a kidney specialist at Seoul National University Hospital and was Dr. Hwang's "right-hand man." Only Dr. Hwang's right-hand man was a woman. Curie is an accomplished scientist in her own right who has made it to the top of her field in a traditionally male-dominated society, and Dr. Lee is much more than the proud owner of a famous dog. He was one of the key members of the research team.

Byeong-Chun was standing before me, beaming.

"Remember me from Washington?" he said, grinning from ear to ear.

"Yes! Snuppy's dad!" I exclaimed. "I wondered if I would get to see you again." Before we knew what was happening, Dr. Lee whisked Bernie and me into a small library, and told us to wait there. Then he left, closing the door behind him.

"This is it," I told Bernie. "We're going to meet Snuppy!"

"You really think so?" I could tell he was about as excited as I was, and from our previous conversations, just as concerned about Snuppy's health.

The next thing we knew, the door popped open and Snuppy came prancing into the room on the end of a leash held by Byeong-Chun. He was instantly recognizable, with his furry, bell-bottomed legs and long, pointed snout. Even as a five-month-old puppy, he possessed the elegance of an Afghan and the reserved personality as well. He greeted us happily but didn't jump all over us the way puppies tend to do. He reminded me of a tall, thin fashion model who knows he's beautiful and doesn't need to win you over. I didn't know if his reserve was something to be concerned about or not, and I could tell that Bernie was wondering the same thing. And of all things, someone had tied a red, white, and blue bandana around his ears and under his chin.

"Why is he wearing a scarf?" Bernie asked with a nervous laugh. "Is it so his ears won't fall off?"

"No . . ." said Byeong-Chun, "it's to keep his ears warm, and to keep them from falling into his food."

I took pictures of Bernie with him, in which Snuppy appeared totally composed, and then I kneeled down to hug him and he gave my face a few licks. Soon Byeong-Chun announced that it was time for Snuppy to go back to his cage. However, when he pulled on the leash, the dog stubbornly plopped down on the tile floor with his front legs spread straight out in front of him. He refused to get up, and it took considerable coaxing to get him onto his feet and to pull him away. If nothing else, he was definitely obstinate. The meeting was all too brief, but then I reminded myself that we were lucky, given the crushing schedule of the last few days, that even a brief meeting with Snuppy had been arranged.

In the taxi en route to our luncheon at a traditional Korean restaurant, Bernie and I swapped observations about Snuppy and our experience at the lab. The lab was impressive for its cutting-edge technology and the obvious industriousness of its large team of researchers. But like me, Bernie, too, thought Snuppy's reserved behavior was unusual for a five-month-old puppy. Everyone we asked assured us that Snuppy was perfectly healthy, but we weren't totally convinced. It simply wasn't consistent with what we knew about other cloned animals.

Over the next few days, the international press hailed the opening of the World Stem Cell Hub as a major step that would greatly accelerate the search for stem cell-based cures. The occasion was highlighted in the world's newspapers and scientific journals as a seminal event, one that set the stage for unprecedented international cooperation. And of course, the Korean press was all over the story. On November 1, the hub opened its website to applicants who wanted to volunteer to donate cells for cloning and for disease research. There were so many applicants from around the world that five minutes after the WSCH started taking online applications, its website crashed.[9] Dr. Hwang's status as a national hero was crystallized. He was living out a fairy tale—a man of humble origins from a small Korean village, who had worked hard, made a major contribution to the well-being of mankind, and been catapulted into the role of a cultural icon. No scientist in recent history had ever enjoyed such star treatment, and the Korean public pinned its vaulted hopes on his continued success in the groundbreaking technology that his research team had pioneered. Gerald Schatten told the *Chosun Ilbo* (one of Korea's most important newspapers), "Professors Hwang Woo-Suk and Ahn Curie of

Seoul National University will be remembered like Pasteur or, indeed, Dr. Curie." With the international spotlight now trained on his every move, Dr. Hwang appeared to be at the summit of a mountain that no one else had ever climbed. And from that summit, only one month later, came perhaps the steepest fall.

chapter ten

korea: the fall

We live in a fallen, tragic world. But we act always with optimism and hope.

—*Laurie Zoloth, Bioethicist, Northwestern University*

Ever since my first visit to Korea, I had been captivated by the carved wooden masks on display at practically every souvenir shop. I learned that they are part of an ancient art form called *Sandaenori*, which has been performed in Korean villages since at least the twelfth century. Carved from the golden wood of the alder tree, the masks are worn by performers in a dramatic dance that tells a story. Their vivid expressions not only represent the personalities of the characters being acted out, but, in keeping with Korea's hierarchical society, they are said to denote the individual's social standing. In elaborate costumed dramas set to drums and music, the sandaenori dances tell stories of satire and sadness, tragedy and mirth. But in order to understand the drama, one has to understand the intricate layering of traditional Korean society, know who's who within the hierarchical order, and interpret the meaning of gestures and movements that serve in the place of words. During my second trip to Korea, it gradually dawned on me that I was being swept up in what felt like a real-life sandaenori.

The first hint of trouble came across the Internet like the distant rumble of thunder. On November 8, 2005, scarcely three weeks after the opening of the hub, the *Korea Times* reported the first suggestion that a storm could be brewing. Korean police were questioning Dr. Sung-il Roh, the director of Mizmedi Women's Hospital, a fertility clinic in Seoul. Roh was suspected of using illegally traded human eggs in fertility treatments at his clinic. As of January 2005, it was illegal in Korea to pay women for egg donations, even if the payment was modest and was restricted to a reimbursement for expenses incurred during the donation process. The *Times* story led with a sentence describing Roh as "a key member of a Korean stem cell team led by professor Hwang Woo-Suk at Seoul National University."[1] It sounded ominous, even though it didn't say directly that Roh had provided illegal eggs to Dr. Hwang's lab. I waited to see if the international news outlets would report on the story, with more detail. But there was a calm before the storm. The story seemed to slip into oblivion, and for several days afterward, it caused barely a ripple.

However, by November 12, there was real trouble brewing. Bernie Siegel called me that morning with the unbelievable news that Gerald Schatten had suddenly—formally and publicly—severed his ties with Woo Suk Hwang. I couldn't believe my ears. The two scientists were not only collaborators but had a celebrated friendship. Schatten was slated to become chairman of the board of directors of the World Stem Cell Hub, which seemed like the opportunity of a lifetime, and this was now clearly impossible. What could have caused Schatten to make such a drastic decision?

Soon, a statement from the University of Pittsburgh was released, in which Schatten said the following: "*I regret to announce that I have suspended my collaborations with Prof. Woo-Suk Hwang, including my involvement with the World Stem Cell Hub project. My decision is grounded solely on my concerns regarding oocyte donations in Dr. Hwang's research reported in 2004* (Hwang et al, *Science* 303, 1669–1674)." This was the celebrated paper in which Dr. Hwang first reported having obtained embryonic stem cells from a cloned embryo, the study that used 242 human eggs in order to derive one stem cell line. Schatten went on to say, "*I continue to believe the scientific accomplishments of Prof. Hwang and his team at Seoul National University, including those in which I had been involved* (Hwang et al, 2005 *Science* 308, 1777–1783; Lee et al, 2005 *Nature* 436, 7051) *are landmark discoveries accelerating biomedical research.*"[2] The second two papers he referred to reported that Hwang had increased the efficiency of therapeutic cloning by at least tenfold, and that he had established 11 patient-specific stem cell lines cloned from the cells of actual patients. This had been lauded as a major breakthrough. It not only showed that human therapeutic cloning is possible, but the reduction in the number of eggs necessary to do it meant that the technology was on its way to becoming feasible on a large-scale basis.

The press release, apparently quoting Schatten, went on to mention that possible "irregularities" in egg donation had been mentioned in May 2004 in *Science* and the scientific journal *Nature*. This was true—allegations had been made that two junior researchers in Dr. Hwang's lab had donated their own oocytes to be used in the experiments—a practice that would not be illegal, but would be frowned upon by international bioethical standards. Having a junior member of the research team make donations at least raises the question of some subtle, or perhaps not-so-subtle pressure, on the women to do so. Both *Nature* and *Science* ran articles printing the claims, but they also reported the fact that Hwang emphatically denied them. Dr. Hwang had not only assured the two journals that the rumors were incorrect, he had apparently convinced his friend and partner, Gerald Schatten, as well.

However, by November 9, 2005, three days prior to his press release, Schatten had obtained different information. He took action immediately: *"Regrettably, yesterday information came to my attention that misrepresentations might have occurred during those oocyte donations . . . compliance concerns with ethical practices for obtaining donated oocytes in their 2004 report, and the resultant breach of trust, are the issues that force me to make this decision."* Astoundingly, the University of Pittsburgh press release wasn't the only missive Schatten had launched—major articles based on interviews he had given to the *Wall Street Journal* and *The Washington Post* were already on American newsstands. Once the articles appeared on the papers' Internet sites, the allegations of unethical practices by the celebrated South Korean team circled the globe, picked up by newspapers, magazines, and television news shows all over the world. Schatten had acted decisively and irrevocably, but most of all, he had made his position known very publicly.

In the first few days after Schatten dropped his bombshell, those of us involved in the stem cell research debates were deeply confused by his precipitous course of action. If he had concerns about the egg donations, and felt that his concerns were serious enough to prompt him to withdraw from his collaboration with Dr. Hwang, why did he feel compelled to do it in such a public way? Couldn't he have quietly informed his partners of his decision, and perhaps requested an investigation by an objective party to confirm all the facts? And what if junior researchers, as was claimed, *had* in fact donated their own eggs? While this was certainly a practice that was best avoided, perhaps the young women had donated eggs out of an altruistic desire to see the research go forward. Schatten's sudden announcement seemed out of proportion to the charges, especially in an embattled field in which opponents of the research would likely seize gleefully on any suggestion of an ethical scandal. Bernie Siegel sent frantic e-mails to both Dr. Schatten and Dr. Hwang, asking for clarification, but neither of them responded. He called Schatten to try to get a better idea of why he taken such a public action, but got no more understanding from talking with him than one could glean from the carefully crafted news release.

Meanwhile, the Korean research team was thrown into a state of disarray. Dr. Hwang was said to be devastated, and Curie Ahn was spearheading an internal investigation on the part of Seoul National University's (SNU's) College of Veterinary Medicine to try to piece together exactly what happened with the egg donations in question. The donations would have taken place in 2002 and 2003, and no one seemed to have any answers about whether or not the claims were true. Stories were bouncing around in the press, and the Koreans seemed to be floundering. Hwang's team was making no statements to the local media and didn't even have a fluent English-speaking representative to communicate with the international press. Dr. Ahn asked for a representative from the Genetics Policy Institute to come to Korea to help with an investigation to establish the facts. They desperately needed an English-speaking representative who was familiar with their work and could help communicate the facts to the foreign press. By the morning of November 15 I was on a flight headed back to Seoul.

Mulling things over on the long flight, I thought that even if the allegations were true, and that Dr. Hwang had misled his collaborator about junior researchers donating their eggs, this was not necessarily fatal to the research itself. While it was entirely possible that the egg donations had occurred in a way that some would find questionable, I knew that they had also occurred before Korea's national regulations of egg donation were in place. Even supposing that the female researchers had given their own eggs, I reasoned, history is full of examples of scientists submitting their own bodies to experimentation out of dedication to their research. And as for payments, it's a fact that in the United States, women are routinely paid a few thousand dollars for egg donations, and the process is perfectly legal. Not optimal in the eyes of many, but certainly not illegal.

Yet, if the statements were true, Dr. Hwang was at serious risk of inviting criticism in a field where hair-trigger sensitivities reigned. I was concerned that the opponents of stem cell research were being handed a reason to lob accusations of unethical behavior at every stem cell researcher in the world. If Dr. Hwang had made mistakes, he needed to come forth immediately, and explain why. If he didn't do so, there could be political fallout for every scientist engaged in the research. But at the same time, Schatten's dramatic decision to sever all ties with him still seemed impulsive and hard to understand. Such an act automatically challenged Dr. Hwang's reputation as a scientist.

After checking in at my hotel, there was no opportunity to rest. I was taken directly to the offices of the World Stem Cell Hub at Seoul National University. It was Wednesday night by then, and I was in a surreal state of sleepless exhaustion. The team of researchers and hospital officials I met up with that night didn't seem much better off. Gathered in the WSCH's brand new conference room, sitting at a table littered with paper cups and styrofoam food containers, they looked like they had been working around the clock for days, trying to untangle the events leading up to the questionable egg donations. Dr. Curie Ahn was leading this group's investigation. She hugged me and welcomed me in, but she looked utterly exhausted.

Curie introduced me to Dr. Myung-Whun Sung, whom she said had been appointed as the media relations person for the hospital. Dr. Sung handed me his card, and I couldn't believe it when I saw the number of positions he held. Not only was he a professor and practicing surgeon at SNU Hospital, he also held positions with the Department of Planning and Budget and the Committee for Global Planning and Development. I was stunned to learn that the university's hospital had no full-time person fielding inquiries and interacting with the press on this scandal. Dr. Sung, a very quiet and serious man, seemed to recognize the gravity of the situation. But I couldn't imagine how one person performing multiple jobs could possibly handle the enormous amount of press activity now being directed at the university, the hospital, the hub, and Dr. Hwang himself.

It was clear to me that night that the Koreans had been blindsided by Schatten's sudden announcement and were struggling to get their bearings while a Korean media tsunami washed over them. Dr. Ahn introduced me to the team that had already begun their investigation, including Dr. Sung-Keun Kang, a close collaborator of Dr. Hwang's at the College of Veterinary Medicine, Sun Ha Paek, who was with SNU Hospital, and Chang-Kyu Lee, a professor at the university's School of Agricultural Biotechnology. That first night, Dr. Sung asked me what I thought of Dr. Schatten's actions, and what people in the United States thought about the case in general. I explained that, like most other people, I didn't know what to think at this stage of the process, but that it was critical for us to piece together all of the facts, and to present them to the world. "But what do *you* think?" he insisted, wanting me to elaborate on my feelings. I paused, thinking he must be asking for my assessment of the damage to Dr. Hwang's research and to the hub. "I don't think it's fatal," I replied. "I think Dr. Hwang can recover from this, if he comes forward with a clear explanation, and assures the world that he will follow the regulations that are in place now." Sung said nothing. It was only later that I realized that what he was asking for was probably not my objective assessment. What he probably wanted from me at that moment was a pledge of absolute loyalty.

Over the next two days, the investigative group met in the hub's conference room and started to piece together what was already known about what happened, when it happened, and what rules were in place when it happened. Members of the team were asked by Dr. Ahn to provide pieces of documentation that would reveal just how the egg donations had been administered. These included the hospital's release statements signed by the women when they agreed to donate, including English translations of the documents for me to review. Dr. Ahn recorded the trajectory of dates and events, including when Dr. Hwang said that he first heard that members of his own research team might have donated eggs (in 2004, according to him), and compared this with his statements to *Science* and *Nature* in which he denied any knowledge of it. SNU's College of Veterinary Medicine's practice of outsourcing human egg retrievals to another hospital, Hanyang Hospital, and Hanyang's outsourcing to Mizmedi Women's Hospital, was laid out and dissected. It was getting easier to see the separation between Dr. Hwang, who is a veterinarian and not qualified to perform human egg retrievals, and a human fertility clinic, which is a completely separate operation. It could appear, from the standpoint of Dr. Hwang's lab, that human eggs were simply being donated anonymously and delivered by the outsource clinic. The women making the donations had supposedly signed confidentiality forms that guaranteed their privacy, which made it seem even more plausible that Dr. Hwang might not have known the identities of the egg donors, unless they went out of their way to tell him.

With Dr. Ahn heading up this investigation, it seemed to me that everything was on the up and up. Despite the fact that everyone taking part in the investigation was clearly tense and upset, for the first few days I felt that SNU's College of Veterinary Medicine was doing exactly what needed to be done—exposing every detail of the egg donations and objectively examining the facts. I felt confident that, even if things looked less than perfect in hindsight, a full disclosure would put the circumstances of the egg donations into perspective.

Dr. Hwang himself attended some of these meetings. He said nothing but was visibly upset. Dr. Ahn told me that he hadn't slept in days, and he looked it. Dr. Hwang was never alone. He was always surrounded by fellow researchers and hospital officials, but at times, he seemed distant, disconnected from what was going on. Above all, he appeared heavily burdened and deeply depressed. Dr. Ahn expressed to me how worried she was about him, even as she seemed to be working herself into the ground trying to untangle everything. Soon she brought Stella Kim, a Korean-born journalist who writes for the *Chicago Tribune*, into the investigation to work with me on English translations.

We all worked straight through Saturday, which is not unusual for the Korean researchers, who often worked seven days a week. On Saturday night, however, just as the story of the egg donations was coming together, I was absolutely shocked when Dr. Ahn announced that she was leaving the country for the next ten days to take care of some business in the United States. She mentioned making visits to San Francisco and San Diego, to talk with some of the hub's U.S. collaborators—presumably to reassure them that the situation would soon be resolved to everyone's satisfaction. Still, knowing how important her friendship was to Dr. Hwang, I couldn't help but wonder if it was a good idea for her to leave the country at such a critical time. After all, even though our investigation was making good headway, Dr. Hwang still hadn't made any statement whatsover to the press, and pressure was building on him to do so. A week had passed since Schatten made his announcement, and reporters from all over the world were asking questions. Dr. Hwang needed to be ready, to have the solid facts at his disposal, but he needed to break his silence as soon as possible. I couldn't believe that Curie Ahn wouldn't be available during this critical period. But the next day, she was gone.

As soon as Dr. Ahn left town, the SNU Hospital's general counsel, an attorney named Sanghan Wang, took charge of the investigation. The entire picture, from my perspective, changed dramatically. The flow of information was suddenly reduced to small bites, and I was no longer privy to the bigger picture. Instead, I had to accept bits of information that were parsed out by the attorney. The whole investigation process was suddenly shrouded in secrecy, and I was forbidden from even communicating with key members of the investigative team, including Dr. Hwang. Most disturbing of all, in place of an open inquiry came a campaign of intimidation.

Mr. Wang, the attorney, sent a formal memo to my attention accusing me of "leaking" information to Stella Kim, the Korean journalist. This was bizarre. Ms. Kim had been brought in by Dr. Ahn herself, openly attended the meetings, and was an active part of the investigation. Suddenly, out of all the people present at the meetings and even a dinner the previous Saturday, Mr. Wang had conducted an "investigation" of my behavior, and found that I had been "sharing information with the press"! It was an unbelievable accusation. And in addition, I was informed by Mr. Wang that Stella Kim had been eliminated from the investigation team and that I was forbidden to communicate with her. I was shocked. Stella was by far the most fluent English speaker I had dealt with, and separating us isolated me. It greatly reduced my access to information uncovered by the investigation. In addition to that, I couldn't imagine why such a competent, hardworking individual would be eliminated as part of the team.

On the heels of this development, I received another memo from Mr. Wang accusing me, essentially, of "impugning the reputation" of the SNU Hospital's attorney in Los Angeles, John Quinn. I almost fell out of my chair when I read it. I had met John Quinn briefly on one occasion only, and knew practically nothing about him or his firm. I couldn't imagine how I could have possibly challenged his reputation or integrity, never mind his firm's. At first I thought the accusation was a simple misunderstanding. When I met Mr. Quinn, I had strongly recommended that if his firm was to select an American public relations firm to assist with media relations in the United States, as they planned to, they should engage one with a track record in handling issues related to the nonprofit, healthcare, and scientific sectors. I couldn't comprehend how this could be interpreted as an attack on Mr. Quinn, his law firm, or their integrity. However, the tone of Mr. Wang's numerous e-mails was truly frightening, and even suggested that I might be in serious legal trouble. I wrote back to him, explaining my position on the matter of the PR firm, and denied that I had ever said one word challenging or impugning the reputation of Mr. Quinn. However, Mr. Wang never really accepted my explanation, and there were several back-and-forth memos belaboring what was essentially a non-issue. After a while, it became clear to me that Mr. Wang didn't really care whether I had actually done any of the things he accused me of. The only thing that mattered was that I admitted some wrongdoing and apologized for it. I was starting to feel like an innocent person being handed a sentence for a crime I didn't commit.

If the goal was intimidation, it worked. After two days of doing almost nothing but fend off accusations from Mr. Wang by e-mail and by phone, I apologized for perhaps expressing my opinion too strongly to Mr. Quinn. But by then the entire picture had shifted, and it was clear that I was dealing with a situation that was a far cry from what I had come to Korea expecting. Instead of welcoming the honest advice and assistance that I assumed was called for, and that Bernie Siegel was also offering from the states, we were informed very clearly that our opinions on how to handle the crisis were not welcome. It was also clear that, rather than continuing their open investigation and sharing the results with the public, the new strategy on the part of the university and hospital was to close ranks, recede behind a wall of secrecy, and let everything be handled by the attorneys.

I was deeply troubled by the new strategy. It seemed obvious that what the situation called for was greater transparency, not secrecy. It seemed I was being intimidated because I had asked too many questions already, and they didn't want me asking any more. And soon the secrecy itself loomed so large that it was starting to seem more and more like an admission of guilt. The more interactions I had with the legal team from the hospital, the more their behavior seemed to suggest that there was more going on than just some confusion about egg donations. As I thought it all through, it occurred to me that there was one, and only one, explanation that would make this new strategy make sense—if there was a *much* bigger problem than improper egg donations, one that involved Hwang's research itself.

On the night of Tuesday, November 22, Korea's Munhwa Broadcasting Corporation (MBC) televised an episode of their documentary series *PD Notebook* that was highly critical of Dr. Hwang and suggested that financially desperate women had been paid by one of Hwang's collaborators for their eggs. Concerns about MBC's motives, and their handling of the situation, had been brewing for weeks. Producers of the show had even gone to the United States and interviewed Korean researchers who had been "seconded," or loaned out, to Gerald Schatten's lab at the University of Pittsburgh to teach their cloning techniques to the U.S. researchers. The SNU researchers complained that the MBC producers were using strong-arm tactics on the scientists to try to get incriminating information about Dr. Hwang. They complained that MBC had been straightforward about their intentions to destroy Dr. Hwang's reputation. MBC even implied that Dr. Hwang was about to be arrested, and if the seconded researchers came clean with MBC, they could avoid serious legal troubles themselves. So everyone knew that at least one negative documentary was forthcoming. However, after airing the first episode of *PD Notebook* that criticized the ethical practices of Hwang's lab, MBC announced that yet another episode was in the works, and they insinuated that it would be even more damaging than the first.

On Thursday, November 24, at 2:00 P.M., Dr. Hwang finally came forward and faced an enormous throng of reporters at a press conference at SNU Hospital. The room was packed tighter than any press event I've ever seen. An exhausted Dr. Hwang walked into the blinding glare of the spotlights and the flaring bulbs of countless cameras. He sat down at a small table, alone, at the front of the room. In the blinding lights, he quickly began to sweat, and his face became flushed. He spoke in Korean, and there was no running translation, so what I saw, perhaps even more vividly, was the raw emotion of the moment. His depression was palpable, and his body language expressed feelings of pure devastation and defeat.

He read from a rather brief statement. At times, he looked up and tears came to his eyes. Each time they did, the clicks of the cameras and the bursts of the flashbulbs exploded. To witness such agony in the glare of such a media frenzy was almost physically painful. It was clear to see that he was accepting guilt—guilt that, at the time, I thought must be a bit overblown. After reading his statement, Dr. Hwang took several questions from the reporters. But as he listened to their questions, some of them quite detailed, a look of shame came over him that was excruciating. Yet, based on what was known about his mistakes up to that point, the apology seemed to me over the top. Here was a man that I had spent hours with, who had shared his research projects with me, who graciously opened up his lab and answered all of my questions. Was it possible that he had seriously deceived me, and deceived the world?

Within the hour, English versions of Dr. Hwang's confession were pouring out in the media. This is what he said: "I am very sorry that I have to tell the public words that are too shameful and horrible. I should be here reporting the successful results of our research, but I'm sorry instead to have to apologize." He went on to describe the events that led up to the accusations of improper egg donations.

Between October 2002 and October 2003, Hwang's research team was attempting to create the first cloned human embryonic stem cells, and they needed a great many human oocytes. As they had admitted all along, the process of cloning at the time was extremely inefficient, requiring literally hundreds of eggs to produce even one blastocyst. But women were not donating enough eggs to keep the research moving as quickly as the scientists wanted it to.

Egg donation is not a simple matter. Women risk considerable side effects from the high doses of hormones taken to put the ovaries into overdrive and induce "super-ovulation." A very small percentage of women have a dangerous reaction to the drugs, which can lead to anything from permanent infertility to—in very rare cases—even death. Complicating the matter is the inconvenience of multiple visits to a hospital or clinic for extensive health screening to qualify for donation, and later, for the egg retrievals themselves. And egg retrieval is no picnic. It involves inserting a needle through the vaginal wall into the adjacent ovaries and extracting the eggs. It's no wonder that egg donations were not keeping pace with the need for them.

The egg shortage was well known in the lab, and according to Dr. Hwang, a female graduate student, who was assisting in the cloning research, approached him, telling him that she wanted to donate her own eggs to the effort. Hwang said that he discouraged her from doing so, because he felt that having a junior member of the research team make such a sacrifice would raise questions of impropriety. Even after he asked her not to do it, the woman approached him two more times, but according to Dr. Hwang, he clearly objected to the idea. In the meantime, another graduate student approached him with the same proposal, and he discouraged her from donating as well. There was nothing illegal about the two donating their eggs, but Dr. Hwang knew that internationally accepted guidelines in medical research would frown upon bodily donations from subordinates in a lab situation. As mentioned earlier, it was understood that, even if they are never directly asked by their superiors, lab subordinates might still feel some subtle pressure or "undue influence" on them to donate.

Eventually according to Hwang, the grad students gave up on trying to win his approval, and they went and made the egg donations without it. According to Hwang, he suspected that the two had gone through with their plans, but they didn't tell him so and he didn't ask. To have asked a young Korean woman such a question, he explained, would violate her sense of modesty, and would have been embarrassing for both of them. So he heard nothing more about the possibility until May of 2004, when David Cyranoski, a writer for the journal *Nature* began investigating rumors that a junior researcher in Hwang's lab had donated her own eggs.

When Cyranoski asked Hwang if the charge was true, he said that he had no knowledge of any lab assistants making egg donations. But because the question had come up, he called Dr. Moon-il Park, who was the chair of Hanyang Hospital's Institutional Review Board (the university's ethical oversight committee). According to Dr. Hwang, Hanyang's rules of confidentiality prohibited him from asking for the exact identities of egg donors, but he asked if any of the donors during the period in question had been researchers from his lab. He told *Nature* that Park had said no.[3] But Dr. Hwang also approached one of the women who had seemed so determined to donate oocytes in the first place. She confessed that she had, but since it was evident that there was growing interest on the part of the media, she begged Dr. Hwang not to reveal it. Dr. Hwang had to make a critical decision. Revealing that a young, unmarried woman had made egg donations could compromise her reputation by Korean standards of honor. She would no longer be seen as virginal and "untouched" by many people in the country. Should he reveal the woman's secret, doing the right thing in the eyes of the scientific community, or should he protect the young woman's desire for privacy?

It wasn't the first time that the brash new world of science had crashed headlong into ancient Korean culture. Hwang decided that, in this case, protecting the women's privacy was the greater virtue. As a matter of fact, every detail of this statement corresponded exactly to what Dr. Hwang told me himself during the investigation, when I sat down with him and asked for his side of the story. Not only that, others had corroborated these facts.

In the end, Dr. Hwang said, "The responsibility for all disputes and controversy lies on me. I will not make any excuse." He resigned from his position as director of the World Stem Cell Hub, but he declared that he would continue to do his research. When he got up to leave, an almost deafening hiss of camera clicks and a blinding explosion of flashbulbs filled the room.

Chivalry, modesty, discretion, and the willingness to sacrifice oneself for others—all are revered values in Korea that weave each individual into the fabric of the collective. Everyone is honor-bound to his family, his employer, his group, and his country. As a foreigner watching it all unfold, I could see that Dr. Hwang's fall from grace was a sorrow and an embarrassment to all Koreans. At the end of the news conference, it was clear that this was a day of national shame for all of Korea. Their Supreme Scientist (a title conferred on him by the government) was disgraced, and their honor was seriously wounded. But their hero had fallen on his own sword in an effort to end everyone's suffering. After the news conference, Dr. Hwang disappeared into seclusion, and I never saw him again. The official story of the hospital was that he had embarked on a period of intense soul-searching at a Buddhist temple.

I returned home to Washington, still deeply confused and concerned about the way the episode was handled. Having seen Hwang's public apology, his level of remorse seemed so far out of proportion to what was confessed in his official statement that the pieces just didn't fit. I could have perhaps chalked that up to Korean culture. However, the intimidation tactics of the attorney had me convinced, in my gut, that Dr. Hwang and his colleagues at the hospital were hiding something *big*.

As it turned out, the sandaenori dance wasn't over, not by a long shot. Only two days before Dr. Hwang's news conference, Sung-il Roh, the doctor at Mizmedi Women's Hospital, had confessed that he *had* been paying women for egg donations, and that some of the eggs had made their way into Dr. Hwang's lab. At his own news conference, Dr. Roh told reporters that Hwang knew nothing of the payments, which he had made out of his own pocket. In fact, he had paid 16 women, including a junior researcher in Hwang's lab, about $1,500 each for the eggs that were used in the groundbreaking work reported in *Science* in 2004. He gave his reason for paying the women as a simple desire to see medical breakthroughs developed for patients with incurable diseases.[4] It wasn't clear whether the payments were restricted to the 16 critical egg donors or if there had been others, but if he had only paid the 16 women and no others, he would have spent $24,000 of his own money. The story added another dimension to Dr. Hwang's troubles, but it also raised the question of whether Dr. Roh truly made the payments out of his own pocket, or if he was being reimbursed and perhaps even turning a profit on the egg donations. At any rate, this happened after Dr. Hwang's press conference, and seemed to reinforce the impression that Dr. Hwang was guilty of no more than being the unwitting recipient of eggs freely donated by his own researchers.

Weave into this intricate cloth an enormous groundswell of public support for Dr. Hwang. Over the following week, at least 1,000 Korean women were said to have contacted the university offering to donate oocytes for his research.[5] On the same day as Dr. Hwang's press conference, the Korean Ministry of Health and Welfare ethics panel officially cleared him of any illegal or unethical activity. They declared that the two researchers had donated their eggs voluntarily, and that Dr. Hwang had acted properly.[6] Seoul National University stood steadfastly by him, and even the Korean president publicly assured Dr. Hwang that the government would continue to support his research and appealed to him to return to work. It quickly became obvious that for the Korean public, this was a patriotic issue, and it wasn't about to relinquish its Hwang-mania.

Munhwa Broadcasting Corporation which had aired the unflattering story about Dr. Hwang, was subjected to a backlash from the public and even from officials in the government. People protested outside the studio and held candlelight vigils for Dr. Hwang. Thousands began boycotting the products of *PD Notebook*'s commercial sponsors, leading all 12 to withdraw advertising from the show. Other news outlets bashed MBC mercilessly, and an "I love Hwang Woo Suk" website logged a tidal wave of complaints about the network's reporting tactics. Threats of violence were even made against employees of the network. Rumors were floating around that MBC had obtained evidence that some of the results published in *Science* regarding the patient-specific cell lines had been faked, but this story was drowned out by an outpouring of anger.

Within a few days, MBC admitted that it had resorted to intimidation in their zeal to obtain information about Dr. Hwang, and made a public apology for violating journalistic ethics. Two producers of *PD Notebook* were suspended and the series itself, after over 14 years on the air, was cancelled.[7]

It seemed that all of this should have lifted Dr. Hwang out of his depression and inspired him to return to his lab. The wind was at his back—the presumed thugs at MBC had been punished, the university and SNU Hospital stood firmly by him, the government absolved him of all blame, the public saw any criticism of him as sacrilege, and the president himself called him back with open arms. If there is any greater validation on earth, it strains the imagination to think of it. It was reasonable to expect Dr. Hwang to emerge any day from his Buddhist retreat, perhaps sober and shaken from the ordeal, but ready to put the incident behind him.

Instead, quite the opposite happened. On December 7, two weeks after the press conference, Dr. Hwang was admitted as a patient at Seoul National University Hospital, apparently in a state of collapse from stress, exhaustion, and possibly a stomach ulcer. I was shocked to see that he was checked into the hospital that was at the epicenter of the events that had caused his stress and exhaustion, the home of the World Stem Cell Hub. It seemed only natural that a person in his position would want to be as far away from those distressing memories as humanly possible. Not only was the story of his hospitalization headlined in every newspaper in Korea, but tragic photos of Dr. Hwang lying in his hospital bed were posted in the newspapers and all over the Internet. In them, his face was pale and he hadn't shaved in days. His eyes were closed in all of the pictures, but his expression was anything but restful, and he looked as though he had lost a considerable amount of weight.

Incredibly, it was clear that the hospital had allowed photographers into Dr. Hwang's room and given them the go-ahead to shoot. The Korean media went wall-to-wall with reports of Dr. Hwang's awful condition. One photo, published in the *Korea Times*, even had the minister of science and technology, Oh Myeong, sitting at Hwang's bedside, holding his hand.[8] The spectacle of his failing health was now the story of the day, and the outpouring of public sympathy only grew more impassioned.

It seemed outrageous for the hospital to allow these photos to be taken, yet it was obviously done with their approval. The drama had taken on a life of its own, driven by some intricate choreography that perhaps no outsider would ever comprehend. Despite the validation and forgiveness poured out for Dr. Hwang, instead of moving on with his life, he seemed to be in a downward spiral. What was the source of his inconsolability? Like a tiny puff of smoke hanging in the air, a small but persistent doubt remained in my mind. It was the question of a much greater wrongdoing than Hwang had already admitted to, a transgression that would require far more public sympathy in order to find forgiveness—the possible doctoring of scientific results.

By December 7, it became clear that I wasn't the only one who suspected that the hospital drama might be a choreographed diversion. In a letter to Chung Un-chan, the president of Seoul National University, a group of SNU professors formally requested an examination of Hwang's data reported in the June 2005 paper in *Science*.[9] This was the paper, coauthored by Gerald Schatten, in which the 11 cloned, patient-specific stem cell lines were described. This was not about improprieties—real or perceived—in egg donations. It was about the actual data. A problem had been discovered by Korean scientists who examined some of the photos of the cell lines submitted to *Science* that were posted on the journal's website. These young scientists noticed that some of the photos were duplicates, when they should have been images of different cell lines. That, in itself, could have been a simple oversight, if someone from the lab had inadvertently sent in the wrong photos. But as they scrutinized the photos further, they observed that in some of them, the DNA "fingerprinting" of supposedly different cell lines was too similar, and they concluded that they could only be the same cell line. This immediately called into question whether the number of cloned cell lines had been exaggerated.[10] *Science* launched a reinvestigation of the data, and asked Dr. Hwang to explain the similarities. The University of Pittsburgh launched its own investigation, while SNU dragged its feet. SNU's initial unwillingness to investigate Dr. Hwang's forensic data—the actual evidence in his lab—was starting to anger other university scientists, who felt that the Hwang affair was starting to reflect badly on the whole Korean research community.

By December 11, the wheels were turning fast. A Seoul-based Internet news site called Pressian released what they claimed was a transcript of an MBC interview with Seon-jong Kim, one of Dr. Hwang's junior researchers now working in Schatten's lab. Kim said that, when preparing the paper for *Science*, Dr. Hwang had asked him to make eleven photos of two stem cell lines, making them all look like different cell lines. Officials at SNU finally sprang into action. They held an emergency meeting and decided that an investigation of Hwang's data was unavoidable.[11]

The next day, Gerald Schatten asked *Science* to remove his name as a coauthor on the landmark 2005 paper that reported the creation of the 11 patient-specific cell lines. The editors at *Science* replied that they had no procedure in place for removing the name of only one author from a previously published paper. In a letter to the journal that was also released to the news wire service Reuters, Dr. Schatten wrote, in a somewhat cryptic style, "My careful re-evaluation of published figures and tables, along with new problematic information, now casts substantial doubts about the paper's accuracy . . . . Over the weekend, I received allegations from someone involved with the experiments that certain elements of the report may be fabricated."[12]

Was it all a case of panic? General hysteria? The whispers of jealous competitors run completely amok? What about martyrdom on the altar of science? The dance went on and on, complete with public demonstrations and more candlelight vigils in support of Dr. Hwang. In Korea, there were many bitter accusations that he had been stabbed in the back by jealous competitors and some even suggested that Schatten had engineered the whole scandal so that he could steal Hwang's discoveries. Would no one represent the voice of reason? Finally, British researchers Ian Wilmut and Keith Campbell came forward with a calm, level-headed solution. "Accusations made in the press about the validity of the experiments published in South Korea are, in our opinion, best resolved within the scientific community," they told Reuters. They noted that questions had been raised when they first announced the news of Dolly. "As we confirmed the validity of our work by cooperating with an independent study, we encourage Hwang's laboratory to cooperate with us to perform an independent test of his cell lines to determine their nuclear and mitochondrial genotype in comparison with the donors of the original cells."[13] The international scientific community immediately weighed in with their support of the suggestion. Perhaps, in the midst of this maelstrom of human foibles, a simple scientific test would shine a ray of light.

On December 12, SNU finally announced that it would conduct an investigation of the evidence in Dr. Hwang's lab. From that day on, the march of increasingly dark events was relentless. For several weeks, stem cell research advocates watched in horror as one disturbing revelation eclipsed another. By December 15, Dr. Hwang's collaborator, Sung-il Roh, who had supplied him with human oocytes, told the Korean media that Hwang had admitted to him that some of the data in the 2005 *Science* paper had been falsified. Both Roh and Hwang asked *Science* to withdraw the paper. Hwang was still in the hospital and still wasn't speaking publicly, but Dr. Roh provided the British newspaper the *Times* with the following explanation: Hwang claimed that his team *had* created the cloned stem cell lines, but nine of the eleven lines died because of a viral infection. Rather than admit this, however, he had them replaced with embryonic stem cell lines created at Dr. Roh's Mizmedi Women's Hospital. Then he had a researcher in his lab manipulate photographs of the cell lines to make it look as though there were more than two lines for the *Science* article. But even this explanation left unanswered questions. First Roh said that the two remaining cell lines had been frozen, and that Dr. Hwang didn't know if they had survived. In the same statement, he said that Hwang told him there were no cloned stem cells in his lab.[14]

To everyone's amazement, on December 16, Dr. Hwang rose from his hospital bed and held another press conference at Seoul National University Hospital. This conference was at least as mobbed with reporters as the last one, if not more so. It was clear that the preceding weeks had taken a toll on him—he was pale and thin, but shaved and dressed in a suit. His statement was nothing short of astonishing. He insisted that his team first created eight cloned stem cell lines, confirming that they were patient-specific cells through DNA tests. Although the cell lines were created in his lab, he did not have the expertise to maintain and grow them, so they were sent to Dr. Roh's lab at Mizmedi Women's Hospital. He also said that he had distributed some of the cloned stem cells "to a handful of universities and research centers in Korea and overseas." In January 2005, calamity struck—six of the genetically tailored cell lines that researchers retrieved from Mizmedi were found to be contaminated, and Dr. Hwang's team could not revive them. Two of the cloned cell lines were left at Mizmedi, frozen, and the researchers didn't know whether they were contaminated or not. Then he claimed that his lab quickly created six more cloned cell lines "to report to *Science*, and three more after that." Even in this statement, it appeared that Hwang had exaggerated the number of lines he reported to *Science*—he was admitting that he had only eight cell lines when he submitted the 2005 paper, but claimed that he derived three more cell lines after doing so.[15]

To the very end, Hwang insisted that he had the ability to clone human embryonic stem cell lines. He said he was "100 percent confident" in his cell lines when he supplied samples of five of them to the MBC TV network along with somatic cells from the patients, for independent verification. But his statement again left unanswered questions. Apparently, cell lines had been moved back and forth between Dr. Hwang's lab at SNU's College of Veterinary Medicine and Dr. Roh's clinic, but it wasn't clear when, how many times, or by whom. At the end of his statement, he claimed that someone who had access to both labs must have switched his cloned cell lines with ordinary embryonic stem cell lines from Dr. Roh's clinic. He called upon the police to investigate the alleged sabotaging of his work. At the very least, his statement included an admission that some of the data in the 2005 paper had been faked. At this point, Dr. Hwang had lost every shred of credibility.

Hwang's admission sent seismic shockwaves through Korean society and through the international scientific community. Korea's medically-related biotechnology stocks took a nosedive, and ordinary Koreans were in such a state of shock that many of them refused to believe their hero had misled them. But in the following days, as several investigations of the matter proceeded, a litany of new problems emerged. In January, *Science* formally retracted both the 2004 paper in which Hwang reported first deriving stem cells from a cloned embryo and the 2005 paper claiming that he had created 11 patient-specific cell lines (while dramatically improving the efficiency of the process). The police closed off Hwang's lab and confiscated computers, notebooks, and lab data. Hwang and a dozen of his key collaborators were forbidden to leave the country, while the police investigated possible criminal wrongdoing. On January 11, 2006, the most damning report yet came from SNU, which reported the findings of its investigation. They announced that the data in Dr. Hwang's ground-breaking papers was fabricated. The report didn't even qualify the matter by saying *some* of the data was faked. Their verdict: "The scientific basis for claiming *any success* [italics are mine] are wholly lacking."[16] Hwang was stripped of his position as chair professor at SNU, and six of his colleagues, including Curie Alan and Dr. Lee (Snuppy's dad), were put on administrative leave.

Just when I thought the whole disaster had bottomed out and things couldn't possibly get any worse, they did. The TV network MBC reported that at least one of Hwang's researchers had been coerced into donating her eggs. And then the awful word "embezzlement" reared its ugly head. The Korean Board of Audit and Inspection concluded that of the approximately $40 million (in U.S. dollars) given to Hwang in 2005 by the government and private investors for research, about $7 million was "inappropriately used."[17] Some had been given to the researchers in Gerald Schatten's lab, in what many alleged to be an attempt to buy their silence over the lack of existing cell lines, and large amounts of money had been "donated" to politicians.[18] Although the amounts varied somewhat in news reports, there could be no question that millions intended for research ended up in one of Hwang's multiple private bank accounts, being used for unauthorized purposes, or were simply unaccounted for.

For the ordinary citizens who had believed fervently in Korea's "Supreme Scientist," the sum of all the revelations was devastating. But out of the wreckage came a tragedy far greater than mere disillusionment with a false hero. For the millions of patients who looked to Dr. Hwang as their brightest hope for a cure, it meant that possibly no patient-specific stem cells had yet been cloned anywhere in the world. They were no closer to a genetically matched stem-cell-based cure than they had been before the rise and fall of the great deception.

The Korean tragedy dealt a painful blow to stem cell researchers the world over. On top of everything, scientists had been led to believe that the technical hurdles of human therapeutic cloning had been crossed by the Koreans. Now it was back to the blackboard for those who had hoped to build on Dr. Hwang's accomplishments. However, in other parts of the world, with far less fanfare and without the intense glare of the spotlight, progress in stem cell research was quietly being made. In the next chapter, I'll show how in Britain, the government is moving steadily toward the realization of stem-cell-based cures for a world that is desperately in need of them.

chapter eleven

winning the peace

To map out a course of action and follow it to an end requires some of the same courage that a soldier needs. Peace has its victories, but it takes brave men and women to win them.

—*Ralph Waldo Emerson*

As tumultuous as the battle over stem cell research is today, most experts believe that the peace will be won when cellular cures begin saving lives. Stem cell research is truly revolutionary, and will no doubt transform the treatment of human disease. The only real questions are when, where, and how this revolution will take place. But to realize its promise, this cutting-edge research needs much more than researchers and labs. It needs a sophisticated level of regulation to optimize the delivery of these innovative cures to the largest number of people worldwide.

Ethical oversight is critical, and so are the safety issues associated with cellular transplants. After all, we expect the drugs that we put into our bodies for medical treatment to be safe, and to meet certain standards for effectiveness. But what about the possibility of integrating into our bodies cells that will become a permanent, self-replicating part of organs and tissues? Shouldn't we know everything we can about those cells before we make the irrevocable decision to introduce them into our bodies?

There is enormous potential for cellular transplants to deliver cures, but their potential for harm is no less dramatic. Embryonic stem cells that have been transplanted into animals in their pluripotent state are well known to cause tumors. Just imagine, for an instant, if you or a loved one received a transplant of multipotent stem cells to cure a disease. Those cells were unknowingly mixed with some pluripotent stem cells that, after transplantation, divided out of control and created a cancerous tumor. You could end up cured of one disease only to develop another. For this reason, scientists are working to understand how to direct the development of embryonic stem cells into either multipotent stem cells or even terminally differentiated cells before they transplant them into people. Transplanted cells could also contain a dangerous virus that could prove fatal to a person who is already sick. These risks must be addressed in a systematic way before stem cell transplants ever live up to their potential for widespread treatments.

There are very good reasons why so many scientists and bioethicists are united in calling for stem cell research to be funded, regulated, and conducted in well-developed countries that have the scientific and regulatory sophistication to oversee it. The tragedy in South Korea has been seen by many as a perfect example of how things can go terribly wrong in a nation without a mature system of oversight. While quite a few skilled scientists are working on stem cell research in Korea, clearly its government did not have an effective system in place to prevent a major scientific fraud, and serious ethical abuses as well. The special issues involved in stem cell research, and the possible ethical abuses in the way it's conducted, call for more advanced oversight than almost any other field, oversight that is simply not available in many countries.

When scientists talk about "oversight" in their field, they mean a complex system of checks and balances. This includes actual laws and regulations, rules and procedures instituted by universities, standards established and overseen by independent professional societies, the supervision of bioethical committees and university review boards, financial and administrative controls, professional standards of conduct, and even a culture of transparency. While a system like this sounds incredibly complicated, it is one that has evolved over decades of scientific development in technologically advanced countries. The system isn't perfect, but through the hard-learned lessons of past mistakes, it works pretty well. Stem cell researchers are especially vocal in calling for such a system to be administered carefully in their field, because in the long run, countless mistakes can be averted that would otherwise only delay the science from going forward.

Of course, there *are* countries, such as Mexico, Russia, and the Ukraine, that are unencumbered by the lengthy process of setting up a centralized system with strict guidelines, where people go today for what are advertised as stem cell transplants. But the patients who seek treatments under these questionable circumstances are putting their lives at risk. As in the case of Susan Fajt, who went to Portugal for treatment, there's a good chance that they are also paying a huge amount of money for a cure that doesn't exist. Today there are even beauty salons in former Soviet countries that promise "stem cell treatments" for skin rejuvenation—a claim that is highly dubious at best, and, with a lack of systematic regulation, could be dangerous as well. It's anybody's guess just what their customers are being injected with, and what it's actually doing to them.

One of the special tasks of administering a national stem cell research program will be to set up stem cell banks. Having a centralized "bank" for stem cell lines is a critical step if treatments to be derived from them are going to be safe and successful. A stem cell bank is an organization that works in a way similar to blood banks today, storing stem cell lines in a frozen state and distributing them to scientists who want to work with them. Most countries where this research is taking place are beginning to address the issue of banking, and today, scientists are talking about the need for international stem cell banks to store and distribute cell lines internationally.

Since President Bush announced his restrictions on stem cell research in 2001, the United States has steadily fallen behind several other countries, such as Britain, Israel, and Singapore, that are rapidly moving ahead in the field. One of the most unfortunate aspects of this situation is the fact that America is one of a handful of countries that is truly equipped to further the field of embryonic stem cell research. It has one of the world's best, if not *the* best, scientific infrastructure, a large pool of research talent, and one of the most developed oversight systems in the world. However, since 2001, the much smaller nation of Great Britain has taken the lead, both in a financial investment that dwarfs that of the U.S. government, and in the establishment of a real system to make the dream of stem cell cures a reality. In 2006, the British government will spend several times the amount that the United States will spend on embryonic stem cell research (the UK plans to spend $177 million, in U.S. dollars,[1] compared to the U.S.'s $38 million).[2]

The lack of government funding in the United States has led to a steady "brain drain" of scientists who have left the country for Britain and other nations where they can do the work that they think is the most promising. For example, in 2005, two prominent cancer researchers, Neal Copeland and Nancy Jenkins, turned down job offers from California's prestigious Stanford University. Even though the passage of Proposition 71 mandates that stem cell research be funded by the state, so far the money is tied up in legal wrangling, meaning that for the time being at least, California researchers are just as hard up for funding as those in most other states. Instead of taking research positions at Stanford, Jenkins and Copeland went to Singapore's Institute of Molecular and Cell Biology, where they could be free to pursue stem cell research with adequate funding from the government.[3] While no one is keeping tabs on the number of scientists leaving the United States, the number is thought to be significant. Dr. Roger Pederson is one U.S. researcher who left the country in 2001 to work at Cambridge University's Stem Cell Institute. He recently told the Bloomberg news agency, "I am here because I think this is the best place to do research using human embryonic stem cells. We're recruiting people constantly from the U.S."[4]

The British government's policy is to support all kinds of stem cell research, and it has led the world in creating the kind of regulatory system that will make sure the research is done ethically, safely, and with the greatest practical benefit. There is no arbitrary cutoff date for the creation of embryonic stem cell lines that can receive research funding, as in the United States. This means British scientists can create new embryonic stem cell lines, using ever-advancing techniques, including the use of therapeutic cloning to create patient-specific cell lines. In addition to that, the UK has already established ethical guidelines to oversee the research and has instituted a material support structure that far exceeds progress in America or any other country.

One of the most important steps toward the development of eventual human treatments has been the establishment of the UK Stem Cell Bank (UKSCB) at the country's National Institute for Biological Standards and Control. The UK Stem Cell Bank, which started to receive funding in January 2003, is considerably ahead of the U.S. National Stem Cell Bank just beginning to take shape. The director of the UKSCB is Glyn Stacey, a microbiologist and cancer researcher who is at the forefront of establishing international standards for how stem cell research, and its resulting treatments, will be developed, monitored, and controlled.

I first met Stacey in South Korea in October 2005, where we both attended the ill-fated opening of the Korean World Stem Cell Hub. As fellow English speakers, Glyn, Bernie Siegel, and I were taken around town together for luncheons, dinners, and one afternoon tour of a Korean palace compound, in which it seemed that we must have walked the entire length of Seoul. Glyn's personal style is one of consummate English understatement, which hides a wry sense of humor. His only, and highly understated, complaint about our round-the-clock marathon was when he approached Bernie and me during the last exhausting leg of our palace tour and plopped down next to us on a park bench with the simple remark, "Death by hospitality."

Because Britain's stem cell bank is the first in the world, I thought that getting a glimpse of how the UKSCB works would be a window into the future for how this entire field could—and probably will—be regulated. I got in touch with Glyn after returning home, and he answered my questions about the UKSCB. I was impressed by the clarity and simplicity of the British system, and how it nevertheless provides a powerful fulcrum for the practical implementation of human therapies from stem cell research.

Stacey has been one of the people drafting the UK standards and guidelines for the cataloguing and testing of stem cell lines. In doing so, he drew upon some of the existing guidelines in human cell and tissue banking, something that he had considerable experience in while working at the UK Centre for Applied Biology and Research in Hertfordshire, England. So far the UKSCB has less than 30 human embryonic stem cell lines, but that number will no doubt grow into the hundreds as more and more cell lines are created. Eventually, the bank will include cell lines derived from adult stems as well, but it doesn't store umbilical cord blood or bone marrow-derived cell lines; there are plenty of other organizations that do that.

Stem cell lines are what Glyn calls "biological medicines." They are biological products, much like transplantable organs and tissues, but in many ways they might be compared to drugs that need to be standardized and tested for safety and effectiveness. One of the challenges of standardization is the fact that stem cell lines are being created in many different labs, being cultivated in a variety of ways, and are even exposed to a diversity of environmental agents. In spite of this, when scientists work with a cell line, and especially when they get to the point of formulating human treatments, they need to know as much about that cell line as possible. If someone were to come up with a cellular "prescription" to treat a diabetic, for example, he would have to know which cell line is the best one to use for that purpose. He would need to know all the therapeutic properties of the line in order to determine how many cells are needed, how effective those cells will be, and whether their effect will be temporary or permanent.

The thinking behind stem cell banks is that scientists who are focused on the cure of disease should not have to devote too many resources into doing what can be standardized by specialists. Well-run cell banks will be instrumental in freeing scientists to do what they do best—search for cures. Not all labs, for example, have advanced expertise in deriving or maintaining cell lines. Some research labs are very good at creating new embryonic stem cell lines, while others are better at driving the differentiation of stem cells into desired cell types. So far, few labs are skilled in maintaining cell lines, which must be stored in a cryogenically frozen state. The way Glyn and many other scientists see an effective system working is that stem cell banks will handle the cataloguing, evaluating, and maintaining of cell lines so that researchers can focus on the development of therapies. The cell banks can also test the cell lines to certify that they are genetically stable and free of infectious agents like viruses and bacteria.

The liberality of British policy means that the UKSCB will eventually have a huge job in regulating a very large field. Because the government provides funding for therapeutic cloning, British scientists could over time create a great many new stem cell lines. As the field matures and more and more scientists work on stem cell-based cures, one wonders how the UKSCB will manage to oversee each and every cell line that is created. It will be able to do this because each newly created embryonic stem cell line in Britain must be licensed.

A license provides a built-in tracking mechanism by requiring that some of the new cells be deposited with the UKSCB for sharing with other scientists in Britain and throughout the rest of the world. The bank will compile the testing and tracking data for all the cell lines, which will be traceable to every scientific project and clinical trial that uses them. In other words, scientists will be able to utilize a safe, well-defined, "off-the-shelf" stem cell line without having to embark on the difficult job of creating it. For example, if a researcher needs cell lines that contain the genes for Alzheimer's disease to be used in the search for a cure, he would be able to obtain suitable cells from the cell bank, without going through the lengthy and difficult process of creating new cell lines for the purpose.

The UKSCB is set up so that it remains free of commercial interests in any of the cell lines it maintains. As for intellectual property rights, the patents on the cell lines in this system remain with the scientists who derived them. For scientists who obtain patented from the cell bank and use them to develop a patentable technique or *cells*, they must negotiate an agreement with the originators of the cell line. Although the UKSCB stores stem cell lines, it does not conduct basic research in stem cell biology with them. This way its assessments and quality control standards for any given stem cell line remain independent of any commercial or professional interest in them.

Stacey is a longtime expert in cell line maintenance and testing proce-dures. He knows cells inside and out. Not only is he director of the UKSCB, he is a professor of cryobiology at the University of Luton, so he's also an expert in freezing and storing them. I asked him what diseases stem cell lines are tested for at the UKSCB, and what kind of health history the bank has on hand about the embryos from which the cell lines are derived. What are the chances, in other words, of someone receiving a stem cell transplant, and the "product" turning out to be defective? It's not as though the bodies of people walking around with these cells can be recalled.

"First of all, we obtain the medical history of the egg and sperm donors from the IVF clinics," he told me. "But we also test the cells for a range of infectious diseases." These include HIV; hepatitis A, B, and C; cytomegalovirus; HTLV (which can cause a crippling demyelination of the spinal cord); and a virus called B19 parvovirus (which, if contracted by a pregnant woman, can cause stillbirth). The cells are also tested for chromosomal abnormalities. Not all cells will be destined for use in human treatments. The UKSCB distinguishes between two grades of cells—research grade and clinical grade. Obviously, the quality and safety threshold is higher for cells that could eventually be used in clinical treatments. Research grade cells won't be tested for B19 parvovirus, for instance, because they will be used for testing drugs or for studying things like cell differentiation, rather than being used in cellular transplants. What about cell lines that are discovered to carry disease genes? "Researchers want them," Stacey said, "to model for different diseases." In other words, they can use the cells to study how specific diseases develop from the very beginning, and the cellular changes that take place along the way. Speaking of modeling for genetic diseases, because the UK allows therapeutic cloning, the bank should soon be housing therapeutically cloned cell lines that express a large number of diseases. These disease-specific cells will be of great use to researchers all over the world who are working on the development of cures.

What the bank will know less about is the presence of genes for late-onset disease in these cells. The egg and sperm donors who use IVF clinics are generally young and healthy, and clinics don't test people for the genes for age-related diseases like heart disease or late-onset Alzheimer's. So it *is* possible that someone could receive a cellular transplant that, 20 or 30 years later, could make them vulnerable to an age-related disease. However, with the bank's ability to keep records over a long period of time, "We will be able to trace the cells from the donor to the final product to the patient," says Stacey. "If there's an adverse effect related to a cell line, we will record it." So the UKSCB not only operates as a clearinghouse and quality control center for stem cell lines, it will be a kind of centralized library compiling massive amounts of information about an enormous number of cell lines, their characteristics, and their end-uses.

This mammoth effort in record keeping, as daunting as it sounds, is critical to operating a large scientific enterprise, one based on "living pharmaceuticals" shared among many people. Stacey and his colleagues must try to envision the enormous number of variables that might enter the picture when these biological medicines enter the stream of human treatments, live in people, interact with the environment, age, change, and eventually die. He has to think about worst-case scenarios, such as infection-contaminated cells or cancer-causing cells making their way into patients. The job is not just one of scientific complexity, it's a job that has enormous administrative challenges.

Quality control, safety testing, cryopreservation, distribution, and record-keeping are not the only functions of the UKSCB. There is also the issue of ethical oversight. The UKSCB's steering committee is charged with making sure that scientists creating new cell lines have followed all of the proper ethical guidelines. These include regulations that cover everything from obtaining the proper consent forms from embryo donors to making sure that the cell lines are used in ethically approved research. While the United States is mired in federal and state-level impediments to therapeutic cloning, the British system already has clear and simple rules in place to regulate it. Cloned embryos may undergo cellular division for only up to 14 days, and then they must be destroyed. (Fourteen days is about the time in the age of an embryo when its cells can begin to differentiate into the parent cells of the three main tissue types.) Egg donation must be voluntary and those accepting eggs must provide women making the donations with information about the risks of taking fertility drugs, and their side effects. The embryos cannot be bought or sold, and they can't be transferred into a woman's body. (If cloned embryos were transferred and resulted in a pregnancy, the offspring would be reproductive clones).

The enormous complexities of designing and administering such a centralized system are why stem cell research experts are calling for the support of governments in the most scientifically developed countries. Quite simply, with government funding comes the ability to regulate. Not only do scientists recognize this imperative, many bioethicists do as well. The colossal shipwreck that was the Korean cloning scandal has been a sobering lesson to the entire world that the highly advanced field of stem cell research needs a regulatory infrastructure that is up to the task of overseeing it. The take-home message behind the rapid rise and fall of Woo Suk Hwang and his now deeply tarnished colleagues is that any attempt to conduct this research without the proper framework of oversight is doomed to collapse.

In the weeks when the Korean scandal was unfolding, opponents of stem cell research were quick to seize on the story as proof-positive that the entire field is unmanageable. They predicted, gleefully, that the whole international enterprise would sink like the Titanic. With characteristic belligerence, Wesley J. Smith, a senior fellow at the conservative Discovery Institute, wrote for the *Weekly Standard*, "Hwang's implosion leaves the field of human cloning research in a state of meltdown. Their poster boy is at best a liar, at worst a fraud and a charlatan who never created human clones at all."[5] And Richard Doerflinger, who is deputy director of the Secretariat for Pro-Life Activities for the U.S. Conference of Catholic Bishops, wrote in the *National Review*, "The fact is that the entire propaganda campaign for research cloning has been filled with misrepresentations, hype and outright lies."[6] But these views came from the highly predictable minority, which stands ready to condemn stem cell research at every turn. As the dust settled, more rational voices emerged.

I spoke with the bioethicist Art Caplan during some of the darkest days of the evolving scandal, when many people were predicting that the enormous magnitude of the South Korean fraud would be a serious blow to the field. Caplan didn't think so. "It may be a short-term setback, but not a long-term one," he said. "Fraud is something that adheres to an individual, but not to a field." As a matter of fact, the story inspired several journalists to look up other famous cases of scientific fraud, such as the legend of the Piltdown Man (the 1912 discovery of the bones of an alleged evolutionary "missing link" in Britain), which turned out to have been fraudulent, and over-reaching, such as the 1989 claims of scientists at the University of Utah that they had achieved cold fusion in a jar of water. Both cases turned out to be false,[7] but they didn't lead to the end of paleontology and physics, respectively. Soon after our conversation, Caplan wrote an article for MSNBC.com in which he pointed out, "Over the years there have been incredible, monstrous frauds perpetrated in geology, paleontology, physics, cancer research, immunology, psychiatry, ophthalmology and psychology. But none—not one—resulted in the end of inquiry or the demise of science."[8]

Instead of inciting a mass exodus of scientists and money from the field, Caplan thinks the more likely effect is that some countries will see the problems in South Korea as a golden opportunity to fill the research void. Scientists who may not have been focused on therapeutic cloning because they believed that the Dr. Hwang had already "cornered the market" once again have an impetus to try to be the first ones to derive human embryonic stem cell lines through nuclear transfer. Caplan doesn't see an end to the international competition for dominance in stem cell research, and there may be lessons in the scandal that will help the field in the long run. The events could have a healthy cautionary effect on those who might be tempted to rush into human clinical trials without doing sufficient animal research first.

I asked Caplan if he thinks stem cell scientists are more prone to ethical corruption than those in other fields, as opponents of the research have claimed. "There's no reason to think that that's true," he said. "But my question is, would this have happened if Hwang were not such a rock star? When he spoke at the University of Pennsylvania last year, I hadn't seen a mob scene like that since Mick Jagger was here."

The story of Hwang's downfall may have more to do with the dangers of celebrity than with the perils of scientific corruption. This is a point that the South Korean media, and the country's people, have examined quite a bit since Hwang's downfall. The Koreans have engaged in a considerable amount of self-criticism over their idolization of Hwang. The phenomenon of over-worshipped celebrities falling from grace is a familiar one to Americans—witness the "crash-and-burn" ruin of numerous Hollywood celebrities, professional athletes, and corporate CEOs. To the never-ending fascination of the rest of us, those who seem to have everything so often plunge over the side of a cliff while grasping for just a little bit more.

In the end, as Caplan and others have pointed out, the scientific system of self-correction worked. Hwang did not get away with his deceptions. In fact, he was outed by young Korean researchers. They are the ones who studied the *Science* website, including all the supplementary data to the two infamous articles, and discovered that something wasn't right. They brought the problems to the attention of the world. "Partly, Hwang was outed as a faker," Caplan wrote for MSNBC.com, "because his colleagues had a sense of integrity. No matter how much fame Hwang attained and no matter how much money the South Korean government threw at Hwang and his team, his colleagues knew he was not forthcoming."[9] He pointed out that it took only seven months from the publishing of the article in which Hwang claimed to have cloned 11 patient-specific stem cell lines until the unmasking of the fraud. Caplan believes that the age of the Internet makes it harder than ever before for scientists to get away with such deceptions. He also believes that all scientific trials should be registered on a website where other scientists can review the data. After all, that is exactly what was Dr. Hwang's undoing.

Even before the Internet, the scientific publishing process has provided a good, though not infallible, system of revealing frauds. But could the editors and reviewers at *Science* have somehow done more to ensure that Hwang's data was authentic? Most scientists say no. Journals don't have the ability to go out and inspect the data of every scientist submitting a paper for publication. They have to trust those supplying the data and assume that it's real, while analyzing the paper critically to see if the whole package appears sound. Every now and then, a paper based on false data might temporarily fall through the cracks. However, time is the enemy of any scientist who wants to commit fraud. Over time, others will try to repeat his experiment, and if there are problems, they will be discovered. Ultimately, faking data is a very dangerous game, and is likely to be a career-destroying move.

Does that mean that no scientist will ever attempt to defraud his colleagues? Of course not. But the percentage of those who try to trick the system is very small. When asked if he thought the peer-review process itself was broken, Donald Kennedy, the editor-in-chief of *Science*, said that in his six years as editor, there had been only three other cases of fraud at the journal.[10] Still, the publishing of new scientific findings calls for dedicated vigilance and the greatest transparency that journal publishers can build into the system. Already the online versions of journals have become a powerful tool in making scientific fakery harder and harder to pull off.

The Korean fraud, in the end, collapsed so resoundingly that it should reassure us that such deceptions, while possible in the short term, are doomed in the end. But what, then, are we to make of Woo Suk Hwang, the man? I have put a lot of thought into this because, when I first met him, there was nothing about him, on the surface, to make me doubt his authority on the subject of cellular cloning. Perhaps the lack of a common language camouflaged what to his own countrymen might have been tip-offs that problems were afoot. But he fooled a lot of people—scores of scientists, diplomats, businessmen, government officials from his own country, and even bioethicists, many of whom visited his lab. And the lab was, without a doubt, an incredibly energetic enterprise and state-of-the-art operation—not, as one might guess, some run-down "front" in a back alley. There was a large team of researchers, if not actually producing the claimed results, working with great skill and apparent dedication toward attaining those goals.

To see Dr. Hwang in his own milieu, he appeared to be in clear command of a large enterprise, and to be just as hardworking and dedicated as his staff. He whisked around the lab, in a great hurry, tolerating no slack or inattention on anyone's part. He was demanding of his subordinates, and it was well known that the researchers in his lab worked long hours, six or seven days a week. All in all, the revelations that he had committed one of the biggest scientific frauds of our time created considerable incongruity for me, and for many others who have met him.

To understand what led Dr. Hwang down the path that eventually snowballed into allegations of a monumental fraud capped by embezzlement would require an exceptional understanding of the vagaries of human nature. But the story, for me, has by now acquired certain outlines by virtue of comparison to similar examples of out-of-control hubris and greed. I am not one of the people who see in Hwang an especially diabolical nature. I think his overall vice has been one of intoxication—by fame, money, and adulation—rather than premeditated evil. He is also a product of a cultural phenomenon in which many, many people colluded and countless others were blinded. In my understanding, he is much like a corporate CEO who "cooked the books" and inflated his organization's stock price in order to keep the investment dollars flowing and the company myth alive. He has more in common with what Kenneth Lay and Jeffrey Skilling of Enron are accused of than he does with any right-wing fantasy of a wild-eyed, evil scientist. In short, the Korean system lacked the layers of oversight that would have stopped Hwang from running his government-funded lab like a private company in which he could do whatever he wanted.

I suspect that in Hwang's universe, the deceptions might have started out small. One thing led to another until the disaster became so huge that literally no one could have averted the crash that dragged dozens of scientists down with him and may yet land him and a few others in prison. Even after he had been discovered in lie after lie, Hwang continued to insist that he had cloned patient-specific stem cell lines, and that he had the technology to reproduce the feat. His allegations that someone in his lab had removed or destroyed the cloned stem cell lines were scoffed at, derided throughout the world because by then, Hwang had lost every shred of credibility. Yet there are reasons to think that if his lab *didn't* clone a single stem cell line, for at least some of the time, he believed that they had. Or, at the very least, he believed that they had the ability, and were on the verge of doing so. There are several reasons why I think this is possible.

Woo Suk Hwang invited prominent scientists from around the world to tour his lab and see his technology. In a lab tour, he might be able to fool a layperson like me, who could be dazzled by such an operation, but it would be lot harder to fool another stem cell research scientist. Yet none of the scientists who toured his lab have publicly expressed any doubts or misgivings about what they saw there. He also loaned, temporarily, three of his cloning experts to work at Gerald Schatten's lab at the University of Pittsburgh, to teach the American researchers the Korean cloning technique. Schatten and his team are experts in animal cloning, and have done extensive work in the field. It seems very unlikely that the visiting scientists could have faked everything, because the American experts would have quickly found them out.

In addition to this, the entire edifice of the World Stem Cell Hub, with the huge investment that went into it, was based on the idea of South Korea sharing its cloning technology through a scientist "exchange system." Korea would send its researchers out to foreign labs to train the scientists there (as they were already doing in Pittsburgh) and they would also bring foreign scientists into the Hwang lab for training. If we assume that the entire enterprise was a fake, that there was no cloning technology and Hwang knew it, surely he would have known that the whole house of cards would collapse in no time at all.

Reinforcing the possibility that Hwang's lab had made improvements in cloning techniques was Snuppy, who was later verified by more than one independent Korean lab to be a clone. Other researchers have tried and failed to clone dogs, but the Hwang team succeeded. And there is one last incident that in my view keeps alive the possibility that Hwang was himself, at least for a time, a believer. That is the way he first handled the investigation conducted by Munhwa Broadcasting Company. When the

TV network questioned the validity of his stem cell lines, Hwang himself gave them samples of the cell lines to be used for independent verification. If he thought that there was any chance of those cell lines not being verifiable, what would have prompted him to hand them over?

Nevertheless, the truth was lost somewhere along the way, and a massive cover-up by Hwang ensued. It's not clear how many people knew the truth that the lab had no patient-specific stem cell lines, because the lab operated on the principle of division of labor, almost like a factory. Everyone had their own piece of the puzzle to work on, and only a few would have seen the big picture. Many Korean analysts have noted that the Confucian hierarchy of Korean culture must have played a major role in dissuading those who knew about the problems from speaking up. In the rigid, authoritarian system of Korean labs, a junior researcher does not question those in authority. I got a strong dose of this maxim myself in my interaction with university and hospital officials. While the investigation of Dr. Hwang was playing out, economist Sang Jo Kim at Hansung University in Seoul told a journalist at *Business Week* about Korean culture, "Lack of trust and integrity, authoritarian culture, and appeals to blind nationalism are problems not limited to the corporate sector. In an organization where whistleblowers are treated as traitors or betrayers, you simply can't stand up against abuse unless you are prepared to risk your whole career."[11] With the university, the hospital, and even the government sticking by Hwang even after numerous deceptions had surfaced, it was easy to see how difficult it would be for any subordinate to blow the whistle.

To complicate matters further, Dr. Hwang became much more than just a star in his field, he became a national symbol of Korea's shining economic future. In this respect, the reliable adage "follow the money" makes it clear that there became a point where enormous sums of money were riding on Hwang's ability to produce cloning technology. After his 2004 announcement that his lab had been the first to derive pluripotent stem cells from a cloned embryo, corporate investors, venture capitalists, university officials, and even politicians flocked to him. Money poured in from the government and from private investors. Hwang and his colleagues had applied for several patents for cloning techniques, and others were eager to get in on the game. Hwang convinced key members of the government and quite a few investors that, with his breakthroughs, Korea would be the world's next biotechnology giant. They would be first in the field to develop, and profit from, some of the most astounding medical technologies of the twenty-first century. Soon, every time Hwang's lab made an announcement, South Korean biotech stocks soared. Hwang was venerated by the media and by rank-and-file Koreans.

There was a flip side to this fairy tale, though, and that was the tremendous pressure to keep producing scientific breakthroughs. The more that was invested in Woo Suk Hwang, the more he had to justify. Eventually, no one could have lived up to the vaunted image the Korean public had of him. But like a star CEO, once he had already overstated his successes, he was compelled to crank out one success after another or his artificially inflated stock prices would sink. The Hwang image, the myth of the unstoppable genius, was after all the lab's stock-in-trade. Somewhere along the way, he had started to cook the books, fudge the data, exaggerate his results. As long as he could continue to demonstrate his brilliance through periodic breakthroughs, by whatever means necessary, he could always go back and fix things later. Hwang might have justified his shortcuts by the age-old business reasoning that, "you don't sell the steak—you sell the sizzle." And Hwang was the sizzle. He was the founding father of the Korean biotech empire.

Toward the end of the debacle, scores of people had their careers, money, and reputations invested in the Hwang "brand name." Even President Roh had thrown his political capital behind him. Many people may have been motivated by greed, vanity, or blind patriotism, but others were swept up into the whirlwind by simple proximity to Hwang. The trouble was that he had been incredibly successful at selling a product that he didn't have. He thought that, with the cloning technology at his fingertips, he would be able to deliver on his promises. But at the moment of reckoning, it didn't matter whether he had the cloning technology; without the cell lines in hand, the center of the enterprise couldn't hold.

However meteoric Hwang's rise to fame and riches was, it was just as short-lived. He was catapulted to international stardom in April 2004 with the announcement that he had produced the world's first bona fide cloned embryo and derived stem cells from it. His fame was heightened in the spring and summer of 2005 when the *Science* article appeared, announcing the alleged cloning of patient-specific stem cell lines, then swiftly followed by the announcement that Snuppy had been born. His celebrity peaked in

October, with the opening of the World Stem Cell Hub, even though by then, behind the scenes, trouble was already brewing. By December 2005, in what must have been one of the most dizzying freefalls a person could ever experience, Hwang had fallen into complete disgrace.

Hwang's downfall was just as public, if not more so, as his scientific career was at its peak. Yet because so many powerful people had their reputations and careers to a great extent dependent on him, many believe that at some point during the investigations he was no longer in control of his actions. He was acting in abeyance to others who might later be able to offer him some rehabilitation. The long, drawn-out hospital drama springs to mind in this respect. Hwang at one time had many powerful friends. As of this writing, it remains to be seen whether any of them are still willing to defend him. But the charges leveled against him of fraud and embezzlement, should he be convicted of them, could land him in prison for up to ten years.

The Korean media received much of the blame for helping to create the phenomenon that was Hwang mania, but in the end they also deserve a great deal of credit for revealing the truth. After all, the MBC network first questioned Hwang's research when he was at the height of his power, and they suffered serious retaliation because of it. And Korean print journalists have been poignant in their soul-searching analyses of what went wrong. Korean publications have deeply criticized the authoritarian tradition and blind patriotism that contributed to the debacle. As the reporter Sang-Hun Choe wrote for the *International Herald Tribune,* "Through Dr. Hwang's fall, South Korea is belatedly learning that biotechnology is not the forum in which to play out its industrial policy ambitions. . . . And the field requires a highly sophisticated regulatory system."[12]

Many people have concluded from the Korean tragedy that stem cell research must proceed slowly. They emphasize, rightly, the imperative of governments to establish ethical guidelines that address the unique issues of stem cell research. One of the less appreciated aspects of the Hwang episode is the need for economic transparency in the administration of research funding—an issue that is critical in countries where government corruption is widespread. The fact that Dr. Hwang found it possible, if not necessary, to bribe government officials to further his work suggests a woeful lack of even the financial checks and balances that are integral to ethical research. These are all good reasons to proceed thoughtfully, but the issues of safety and the just distribution of the benefits of stem cell research call for our attention as well. All of these issues should cause us to pause and carefully consider the possibilities of abuse in science. But they should not lead us into a state of paralysis.

The South Korean tragedy only highlights the fact that it is incumbent on the most scientifically advanced nations, such as the United States and Britain, to lead the way in supporting stem cell research. These countries already have sophisticated mechanisms of oversight that can be modified and adapted to address the special issues of stem cell research. They are qualified to formulate a regulatory system that might become international in scope, ensuring that the research is conducted properly throughout the world. These countries have universities that can provide additional ethical oversight and scientific communities that are willing and fully able to launch the field of stem cell research. They also have developed economies that will lead to the implementation of cures—the biotech companies that can turn basic research into therapies and deliver those therapies into the hands of doctors. Of course, even in this scenario, there will be problems and challenges to overcome, but at least there are rules to play by.

Much to the frustration of most Americans, the United States has virtually abdicated its responsibility to take a leading role in the development of this new science. Our government of the past five years has put the interests of millions of patients behind its desire to appease a highly vocal political and religious minority—in a word, it has allowed politics to trump the public good. This influential minority wants medical research to come to a standstill while they argue ad infinitum over ideological abstractions that will probably never enjoy a consensus. Meanwhile, millions of men, women, and children suffer and die. If Americans can't turn the tide on this paralyzing, anti-progressive trend, we should be prepared to accept a future of economic decline and second-world status, and with the aging of the U.S. population, widespread illness and disability.

To those who point out that stem cell science, or any other science, could possibly be abused, I say that this is no reason to abandon its enormous potential for good. After all, in 2001, terrorists flew airplanes into the World Trade Center in New York and killed thousands of people, but we haven't outlawed airplanes as a result. In the last analysis, no one can regulate integrity. The idea of placing a "freeze" on progress because there are good and bad people in the world, because knowledge can be misused, or because we can't always guarantee the outcome is an assault on the human spirit. It is living by our worst fears, not by our greatest hopes.

Furthermore, the imposition of extreme religious beliefs, shared by a minority and forced upon the majority, is unworthy of a democratic society. No matter how much American religious conservatives want to impose a state of arrested scientific development in their own country, new discoveries and technologies will proceed in other parts of the world. The only possible outcome of the suppression of science in the United States is to diminish our influence in a world that depends increasingly on science and technology, while leaving powerful technologies in the hands of others.

If anyone objects to receiving a stem cell treatment on the basis of their religion, I want to emphasize that they have they have every right to do so. But they don't have the right to withhold that treatment from others who are sick and suffering from cancer, diabetes, heart disease, stroke, paralysis, neurological disease, Parkinson's, Alzheimer's, multiple sclerosis, muscular dystrophy, ALS, osteoarthritis, severe burns, brain damage, cystic fibrosis, lung disease, kidney and liver disease, congenital birth defects, or a host of other conditions.

Biomedical discoveries will continue to grow because they are driven by one of the most compelling needs of all time—to save lives and relieve human suffering. Our religious and cultural traditions, as valuable as they are, must be enlarged in light of the biological revolution. There is no intrinsic reason, in my mind, for religion to be at odds with science, unless it refuses to adapt to the expanding truth of human progress. At the same time, science isn't likely to ever answer the big questions that fall into the domain of religion. It will never "cure" the human condition. But it can certainly improve it.

The question is, will we have more compassion for theoretical, potential persons than we do for the living? Our most urgent need is to save human lives; our most enduring one is to find meaning in them. We should trust ourselves to identify the core values of our religious traditions and separate them from the transient scientific and social understandings at the time of their inception. It is now the great challenge of our religious and philosophical traditions to adapt to modern-day realities, and to bring their best wisdom to bear on some of the most important decisions we will ever make.

notes

chapter one

1. "Life Expectancy at Birth, at 65 Years of Age, and at 75 Years of Age, According to Race and Sex: United States, Selected Years 1900–2001," National Center for Health Statistics, 2003. www.cdc.gov/nchs/data/hus/tables/2003/03hus027.pdf (accessed July 14, 2003).
2. Alzheimer's Association, "About Alzheimers: Causes and Risk Factors," 2006. http://www.alz.org/AboutAD/causes.asp (accessed February 27, 2006).
3. DEVCAN: *Probability of Developing or Dying of Cancer Software*, version 5.2, Statistical Research and Applications Branch, National Cancer Institute, 2004. http://srab cancer/gov/devcan (accessed May 20, 2005).

chapter two

1. Cynthia Tucker, "American Know-How Hobbled by Know-Nothings," Working for Change, http://www.workingforchange.com/printitem.cfm?itemid = 19455 (accessed April 25, 2006).
2. The Trustees of Princeton University, "The 2004 Jurow Lecture: From Recombinant DNA to Stem Cells—Making Science Policy in a Democracy," by Shirley M. Tilghman. www.princeton.edu/president/speeches/2004331/index.xml (accessed December 19, 2005).
3. h2g2, "The History of Modern Medicine," BBC, October 1, 2001. http://www.bbc.co.uk/dna/h2g2/A600418 (accessed December 20, 2005).
4. A. J. Maas, "General Resurrection," New Advent Catholic Encyclopedia, October 6, 2005. http://www.newadvent.org/cathen/12792a.htm (accessed February 22, 2006).
5. Eve Herold, "Update on Stem Cell Research, Editorial," Stem Cell Research Foundation, April 2002.
6. h2g2, "The History of Modern Medicine."
7. Andrew Dickson White, "Theological Opposition to Anatomical Studies," *A History of the Warfare of Science with Theology in Christendom* (New York: D. Appleton and Company, 1896; 1898). http://www.cscs.umich.edu/-crshalizi/White/ (accessed February 27, 2006).
8. Religious Tolerance, "Religious Change and Past Religious Conflicts," http://www.religioustolerance.org/past_mor.htm (accessed December 15, 2005).
9. White, *A History of the Warfare of Science with Theology in Christendom*.
10. Richard Gordon, *The Alarming History of Medicine*, (New York: St. Martin's Press, 1993), p. 83.
11. Gift of Life Donor Program, "Donation and Transplantation History." http://www.donors1.org/donation/history.html (accessed December 19, 2005).
12. Louisa Moon, "Historical Milestones in Transplantation," 2002. http://www.miracosta.edu/home/1moon/HistoryOT.html (accessed December 19, 2005); Maeve Haldane,

"Cultural Concepts of Brain Death and Transplants, *McGill Reporter* 34, no. 9 (January 24, 2002). http://www.mcgill.ca/reporter/34/09/lock/ (accessed December 19, 2005).
13. Catholic Insight, "What Is the Catholic Position on In-Vitro Fertilization (IVF)?," January 27, 2006. http://catholicinsight.com/online/church/vatican/article_476.shtml (accessed February 27, 2006).
14. Robert Edwards and Patrick Steptoe, *A Matter of Life: The Story of a Medical Breakthrough*, (New York: William Morrow and Company, Inc., 1980), pp. 141–142.
15. Edwards and Steptoe, *A Matter of Life: The Story of a Medical Breakthrough*, pp. 146–155.
16. David I. Hoffman, Gail L. Zellman, C. Christine Fair, Jacob F. Mayer, Joyce G. Zeitz, William E. Gibbons, and Thomas G. Turner, "Cryopreserved Embryos in the United States and Their Availability for Research," *Fertility and Sterility*, 79, no. 5 (May 2003).
17. Beliefnet, "Discarding of Embryos Fuels Debate on Stem Cell Research." http://www.beliefnet.org/story/152/story_15206.html (accessed February 27, 2006).

chapter three

1. Yuehua Jiang, Balkrishna N. Jahagirdar, R. Lee Reinhardt, Robert E. Schwartz, C. Dirk Keene, Xilma R. Ortiz-Gonzalez, Morayma Reyes, et al., "Pluripotency of Mesenchymal Stem Cell Derived from Adult Marrow," *Nature* 418 (2002): 41–49.
2. Gareth Cook, "From Adult Cells Comes Debate," *Boston Globe*, November 1, 2004, p. 5.
3. Cook, "From Adult Cells Comes Debate."
4. Reuters, "Adult Stem Cells Transfer Improves Heart Function," July 8, 2004; Reuters, "Bone Marrow Cells Help Heart Failure in Experiment," January 25, 2005.
5. Stanford School of Medicine, Office of Communication and Public Affairs, "Stanford Q&A: Irving Weissman on the South Korea Stem Cell Controversy." http:// mednews. stanford.edu/releases/2005/december/stemcell5ques.html (accessed January 1, 2006).
6. Alexandra Goho, "Stem Cells Enable Paralysed Rats to Walk," *NewScientist.com*, July 3, 2003. http://www.newscientist.com/article.ns?id = dn3894&print = true (accessed July 8, 2003).
7. J. A. Thomson, J. Itskovitz-Eldor, S. S. Sahpiro, M. A. Waknits, J. J. Swiergiel, V. S. Marshall, and J. M. Jones, "Embryonic Stem Cell Lines Derived from Human Blastocysts," *Science*, 282 (1998): 1145–1147.
8. Michael J. Shamblott, Joyce Axelman, Shunping Wang, Elizabeth M. Bugg, John W. Littlefield, Peter J. Donovan, Paul D. Blumenthal, George R. Huggins, and John D. Gearhart, "Derivations of Pluripotent Stem Cells from Human Primordial Germ Cells," *Proceedings of the National Academy of Sciences,*. 95, no. 23 (1998): 13726–13731.
9. S-Chun Zhang, Marius Wernig, Ian D. Duncan, Oliver Brustle, and James A. Thomson, "In Vitro Differentiation of Transplantable Neural Precursors from Human Embryonic Stem Cells," *Nature Biotechnology*, 19 (2001): 1129–1133; Benjamin E. Reubinoff, Pavel Itsykson, Tikva Turetsky, Martin F. Pera, Etti Reinhartz, Anna Itzik, and Tamir Ben-Hur, "Neural Progenitors from Human Embryonic Stem Cells," *Nature Biotechnology* 19 (2001): 1134–1140.
10. Lars M. Björklund, Rosario Sánchez-Pernaute, Sangmi Chung, Therese Andersson, Iris Yin Ching Chen, Kevin St. P. McNaught, Anna-Liisa Brownell, Bruce G. Jenkins, Claes Wahlestedt, Kwang-Soo Kim, and Ole Isacson. "Embryonic Stem Cells Develop into Functional Dopaminergic Neurons After Transplantation in a Parkinson Rat Model," *Proceedings of the National Academy of Sciences* 99 (2002): 2344–2349. http://www.pnas.org/cgi/content/full/99/4/2344 (accessed April 17, 2006).

11. J. B. Cibelli, R. P. Lanza, M. D. West, C. Ezzell, "The First Human Cloned Embryo," Scientific American 286, no. 1 (2002): 44–51.

12. Jose B. Cibelli, Kathleen A. Grant, Karen B. Chapman, Kerrianne Cunniff, Travis Worst, Heather L. Green, Stephen J. Walker, et al., "Parthenogenetic Stem Cells in Nonhuman Primates," Science 295 (2002): 779–780.

13. W. M. Rideout III, K. Hochedlinger, M. Kyba, G. Q. Daley and R. Jaenisch, "Correction of a Genetic Defect by Nuclear Transplantation and Combined Cell and Gene Therapy," Cell 109 (2002): 17–27.

14. Benjamin Dekel, Ninette Amariglio, Naftali Kaminski, Arnon Schwartz, Elinor Goshen, Fabian D. Arditti, Ilan Tsarfaty, Justen H. Passwell, Yair Reisner, and Gideon Rechavi, "Engraftment and Differentiation of Human Metanephroi into Functional Mature Nephrons after Transplantation into Mice is Accompanied by a Profile of Gene Expression Similar to Normal Human Kidney Development," Journal of the American Society of Nephrology 13 (2002): 977–990.

15. Douglas A. Kerr, Jerònia Lladó, Michael J. Shamblott, Nicholas J. Maragakis, David N. Irani, Thomas O. Crawford, Chitra Krishnan, Sonny Dike, John D. Gearhart, and Jeffrey D. Rothstein, "Human Embryonic Germ Cell Derivatives Facilitate Motor Recovery of Rats with Diffuse Motor Neuron Injury," The Journal of Neuroscience 23, no. 12 (2003): 5131–5140.

16. Qilin Cao, Xiao-Ming Xu, William H. DeVries, Gaby U. Enzmann, Peipei Ping, Pantelis Tsoulfas, Patrick M. Wood, Mary Bartlett Bunge, and Scott R. Whittemore, "Functional Recovery in Traumatic Spinal Cord Injury after Transplantation of Multineurotrophin-Expressing Glial-Restricted Precursor Cells," The Journal of Neuroscience 25, no. 30 (2005): 6947–6957.

17. Shulamit Levenberg, Ngan F. Huang, Erin Lavik, Arlin B. Rogers, Joseph Itskovitz-Eldor, and Robert Langer, "Differentiation of Human Embryonic Stem Cells on Three-Dimensional Polymer Scaffolds," Proceedings of the National Academy of Sciences 100, no. 22 (2003): 12741–12746.

18. Hanna Segev, Bettina Fishman, Anna Ziskind, Margarita Shulman, and Joseph Itskovitz-Eldor, "Differentiation of Human Embryonic Stem Cells into Insulin-Producing Clusters," Stem Cells 22 (2004): 265–274.

19. Izhak Kehat, Leonid Khimovich, Oren Caspi, Amira Gepstein, Rona Shofti, Gil Arbel, Irit Huber, Jonathan Satin, Joseph Itskovitz-Eldor, and Lior Gepstein, "Electromechanical Integration of Cardiomyocytes Derived from Human Embryonic Stem Cells," Nature Biotechnology 22, (2004):1282–1289.

20. Yasushi Takagi, Jun Takahashi, Hidemoto Saiki, Asuka Morizane, Takuya Hayashi, Yo Kishi, Hitoshi Fukuda, "Dopaminergic Neurons Generated from Monkey Embryonic Stem Cells Function in a Parkinson Primate Model," Journal of Clinical Investigation 115 (2005): 102–109.

21. Holly Wagner, "Researchers Devise Ways to Mass-Produce Embryonic Stem Cells," Ohio State University website, March 15, 2005. http://researchnews.osu.edu/archive/masstem. htm (accessed March 17, 2005).

22. Sandra M. Klein, Soshana Behrstock, Jacalyn McHugh, Kristin Hoffman, Kyle Wallace, Masatoshi Suzuki, Patrick Aebischer, Clive N. Svendsen, "GDNF Delivery Using Human Neural Progenitor Cells in Rat Model of ALS," Human Gene Therapy 16, no. 4 (2005): 509–521.

23. "Stem Cell Researchers Create Eggs," *Sydney Morning Herald*, June 20, 2005. www.smh. com.au/news/science/stem-cell-researchers-create-eggs/2005/06/20/11192 (accessed June 21, 2005).

24. Petter S. Woll, Colin H. Martin, Jeffrey S. Miller, and Dan S. Kaufman, "Human Embryonic Stem Cell-Derived NK Cells Acquire Functional Receptors and Cytolytic Activity," *The Journal of Immunology* 175 (2005): 5095–5103.

25. Joanne Morrison, "Stem Cell Work in U.S. Takes Big Leap: Lab Grew Line Free of Animal Product," Reuters, January 2, 2006. www.boston.com/news/nation/articles/2006/01/02/stem_cell_work_in_us_takes_big_ (accessed February 1, 2006).

chapter four

1. American Association for the Advancement of Science, "House Approves NIH Budget with 2.6 Percent Increase in 2005," July 28, 2004 (revised September 9, 2004). http://www.aaas.org/spp/rd/proj05p.htm (accessed December 11, 2005).

2. U.S. Department of Health and Human Services, National Institutes of Health, "Estimates of Funding for Various Diseases, Conditions, Research Areas" (updated February 3, 2006). http://www.nih.gov/news/fundingresearchareas.htm (accessed April 4, 2006).

3. Howard Hughes Medical Institute, "New Embryonic Stem Cell Lines to Be Made Available to Researchers," March 3, 2004. http://www.hhmi.org/news/melton4.html (accessed February 27, 2006).

4. Opinion Research Corporation, "American Views on Stem Cell Research in the Wake of the Death of Ronald Reagan," prepared for Results for America, June 16, 2004.

5. Ramesh Ponnuru, "Hard Cell," *National Review Online*, June 4, 2004. http://www.nationalreview.com/ponnuru/ponnuru200406040842.asp (accessed June 10, 2004).

6. Jennifer Warner, WebMD Medical News, "Stem Cell Support Rising After Reagan's Death," June 16, 2004. http://my.webmd.com/content/Article/88/ 100108.htm? printing = true (accessed June 17, 2004).

7. Republican Main Street Partnership, "New GOP Poll and Ad Campaign Boost Republican Main Street Partnership Support for Expanded Federal Stem Cell Policy," news release, May 12, 2005.

8. American Bioethics Advisory Commission, Mission. http://www.all.org/abac/mission. htm (accessed December 12, 2005).

9. Kathryn Hinsch, "Bioethics and Public Policy: Conservative Dominance in the Current Landscape," Women's Bioethics Project report, November 2005.

10. Aubrey Noelle Stimola, "Q. for Bush: If Embryonic Stem Cell Work is Stopped, Should IVF Be Next?" Health Facts and Fears.com, May 24, 2005. http://www.acsh.org/print Version/hfaf_printNews.asp?newsID = 558 (accessed May 24, 2005).

11. National Child Welfare Resource Center for Adoption, 2006. http://nrcadoption.org/ (accessed January 10, 2006).

12. Ralph Nader, *The Good Fight: Declare Your Independence and Close the Democracy Gap* (New York: HarperCollins, 2004), 54.

13. Peter Samuelson, interview by Raney Aaronson, *Frontline*, PBS, November 8, 2005.

14. University of Kansas Medical Center, "Stem Cell Basics: Legislators Toolkit: Federal Public Policy," 2005. http://www.kumc.edu/stemcell/toolkit4.html (accessed February 26, 2006).

15. *Human Cloning Prohibition Act of 2005*, S 658 IS, 109th Cong., 1st sess. (March 17, 2005).

16. Lynda Richardson, "The Slippery Intersection of Medicine and Politics," *New York Times*, July 27, 2004, B2.

chapter five

1. The White House, "President Delivers State of the Union Address," news release, January 29, 2002.
2. U.S. Department of State, Bureau of Public Affairs, "To Ban Human Cloning," 2002.
3. Catholic Online, "Holy Seeks Call for a Ban on All Human Cloning," September 30, 2003. http://www.catholic.org/ printer_friendly.php?id = 385&section = Featured = Today (accessed January 18, 2006).
4. Kate Dewes, "Taking Nuclear Weapons to Court," Disarmament and Security Centre. http://disarmsecure.org/publications/papers/court.html (accessed January 18, 2006).
5. CBS News.com, "U.N. Puts Cloning Talks on Hold," December 9, 2003. http://www.cbsnews.com/stories/2003/12/09/tech/printable587616.shtml (accessed February 27, 2006).
6. Gregory M. Lamb, "U.N. Delay: A Boost for Cloning Advocates," *Christian Science Monitor*, October 25, 2004, A12.
7. U.N. News Centre, "U.N. Committee Approves International Declaration Against Human Cloning," news release, February 21, 2005.

chapter seven

1. Tim Townsend, "Thinking on Abortion, Fetus, Guides Most Groups' Positions," *St. Louis Post-Dispatch*, August 22, 2004. http://www.stltoday.com/stltoday/emaf.nsf/Popup? ReadForm&db = stltoday%5Cnews%5C (accessed August 23, 2004).
2. Michael Kinsley, "The False Controversy of Stem Cells," *Time*, May 31, 2004.
3. Stephen S. Hall, "The Good Egg," *Discover*, May 2004.
4. Harold J. Morowitz and James S. Trefil, *The Facts of Life: Science and the Abortion Controversy* (Oxford, Eng.: Oxford University Press, 1992), p. 51.
5. Judith A. Johnson and Erin D. Williams, "CRS Report for Congress: Stem Cell Research," Congressional Research Service, Library of Congress, May 20, 2005.
6. Snowflakes Embryo Adoption Program, Embryo Adoption Awareness Campaign, "Legislative Homework." http://www.embryoadoption.org/aboutlegframework.asp (accessed February 23, 2006).
7. Louis M. Guenin, "Morals and Primordials," *Science* 292 (2001): 1659–1660.
8. Jane Maienschein, *Whose View of Life?* (Cambridge, MA: Harvard University Press, 2003), pp. 25–26.
9. Jennifer Skalka, "Churches to Weigh in on Stem Cell Research," *Baltimore Sun*, January 24, 2006. http://www.baltimoresun.com/news/health/bal-stem0124,4620757.story?coll = bal-home (accessed January 25, 2006).
10. Skalka, "Churches to Weigh in on Stem Cell Research."
11. Alan G. Padgett, "On the Ethics of Stem Cell Research," *Science and Theology News*, July 13, 2005. http://www.stnews.org/print.php?article_id = 994 (accessed January 27, 2006).
12. Yoel Jakobovits, "Judaism and Stem Cell Research," Torah.org. http://www.torah.org/features/secondlook/stemcell.html?print = 1 (accessed July 15, 2005).
13. Michele Weckerly, "The Islamic View on Stem Cell Research," *Rutgers Journal of Law and Religion*, September 30, 2002. http://www-camlaw.rutgers.edu/publications/law-religion/Dnew02.htm (accessed February 10, 2006).

14. Steven Zecola, "President Bush Misleads the Public about Stem Cell Research," *PRWeb*, September 9, 2004 (accessed February 26, 2006). http://www.prwebdirect.com/releases/2004/10/prweb166767.htm.
15. Damien Keown, "No Clear Buddhist Stance on Stem Cell Work," *Science and Theology News*, April 1, 2004. http://www.stnews.org/print.php?article_id = 860 (accessed January 27, 2006).
16. Townsend, "Thinking on Abortion, Fetus, Guides Most Groups' Positions."
17. M. Elkin, "Church Announces Campaign Against Stem Cell Research," International News Alliance, inadaily.com, March 31, 2005. http://www.iht.com/getina/files/236149.html (accessed March 31, 2005).
18. H. J. Landry and L. G. Keith, "The Vanishing Twin: A Review," *Human Reproduction Update*, 4, no. 2 (1998): pp. 177–183.
19. Claire Ainsworth, "The Stranger Within," *New Scientist*, 180, no. 2421 (2003): p. 34. http://www.katewerk.com/chimera.html (accessed February 26, 2006).
20. Gene Outka, "The Ethics of Human Stem Cell Research," 29–64, *God and the Embryo: Religious Voices on Stem Cells and Cloning*, ed. Brent Waters and Ronald Cole-Turner (Washington, DC: Georgetown University Press, 2003), p. 47.
21. Wesley J. Smith, "The 'Wrong' Cure," *National Review Online*, September 9, 2004. http://www.nationalreview.com/script/printpage.asp?ref = /smithw/smith20040909083 5.asp (accessed September 13, 2004).
22. Joseph M. Incandela, "The Catholic Church and Surrogate Motherhood." surrogacy.com. http://www.surrogacy.com/religion/catholic.html (accessed February 11, 2006).

chapter eight

1. David A. Shaywitz, "Stem Cell Hype and Hope," *Washington Post*, January 12, 2006.
2. Aaron Levine, "Trends in the Geographic Distribution of Embryonic Stem Cell Research," *Politics and the Life Sciences*, 23, no. 2 (2005): 40–45.
3. National Family Caregivers Association, "Family Caregivers—America's Invisible Workforce of 50 Million," news release, October 8, 2004.
4. Jeanie Lerche Davis, "Medical Bills Can Lead to Bankruptcy," WebMD Medical News, February 2, 2005. http://my.webmd.com/content/Article/100/105540.htm? printing = true (accessed February 10, 2005).
5. CNN.com, "Percent Who Skip Health Insurance Jumps," http://cnn.worldnews. printthis.clickability.com/pt/cpt?action = cpt&title = CNN.com+-+Perc (accessed April 26, 2006).
6. Ibid.
7. Marc Kaufman and Rob Stein, "Record Share of Economy Spent on Health Care," *Washington Post*, January 10, 2006.
8. Center on Budget and Policy Priorities, "The Number of Uninsured Americans Continued to Rise in 2004, August 30, 2004. http://www.cbpp.org/8–30–05health.htm (accessed February 26, 2006).

chapter nine

1. BBC News, "Fresh Doubt on Pig Organ Safety," August 16, 2000. http://news.bbc.co.uk.1/hi/health/883359.stm (accessed December 1, 2005).
2. Tae-gyu Kim, "Stem Cell Clinical Test Facility Launched," *Korea Times*, November 7, 2005. http://hankooki.com/service/print/Print.php?po = times.hankooki.com/1page/200511/k (accessed November 7, 2005).

3. The Welcome Trust, "Genetic Modification of Pigs for Xenotransplantation," July 30, 2003. http://www.wellcome.ac.uk/en/genome/tacklingdisease/hg09f001.html (accessed December 1, 2005).

4. Woo Suk Hwang, Young June Ryu, John Hyuk Park, Eul Soon Park, Eu Gene Lee, Ja Min Koo, Hyun Yong Jeon, et al., "Evidence of Pluripotent Human Embryonic Stem Cell Line Derived from a Cloned Blastocyst," *Science* 303 (2004): 1669–1674.

5. Woo Suk Hwang, Sung Il Roh, Byeong Chun Lee, Sung Keun Kang, Dae Kee Kwon, Sue Kim, Sun Jong Kim, et al., "Patient-Specific Embryonic Stem Cells Derived from Human SCNT Blastocysts," *Science* 308 (2005): 1777–1783.

6. Rowan Hooper, "World's First Canine Clone is Revealed," NewScientist.com, August 3, 2005. http://www.newscientist.com/article.ns?id = dn7785&print = true (accessed February 26, 2006).

7. Calvin Simerly, Tanja Dominko, Christopher Navara, Christopher Payne, Saverio Capuano, Gabriella Gosman, Kowit-Yu Chong, et al., "Molecular Correlates of Primate Nuclear Transfer Failures," *Science* 300 (2003): 297.

8. Yoo, Fairneny, and Associates, "Korean Legal Codes for Biomedical Ethics and Safety— Code 7150," English translation, 2005. www.sciencemag.org/cgi/dara/112286/DC1/1.

9. Digital Chosun, "Hope for Stem Cell Cure Draws Huge Crowds," *The Chosun Ilbo*, November 1, 2005. http://english.chosu.com/cgi-bin/printNews?id = 200511010023 (accessed November 1, 2005).

chapter ten

1. Tae-gyu Kim, "Member of Hwang Team Involved in Ovum Scandal," *Korea Times*, November 8, 2005. http://times.hankooki.com/service/print/Print.php?po = times.hankooki.com/lpage/200511/k (accessed November 8, 2005).

2. University of Pittsburgh, "University of Pittsburgh Researcher Ends Collaboration With South Korean Stem Cell Program," news release, November 12, 2005.

3. David Cyranoski, "Stem Cell Pioneer Resigns," News@Nature.com, November 24, 2005. http://www.nature.com/news/2005/051121/pf/051121–12_pf.html (accessed November 24, 2005).

4. Ji Soo Kim, "Korean Stem-Cell Researcher Under Fire," Ohmy News.com, November 22, 2005. http://english.ohmynews.com/articlereview/article_print.asp?menu = c10400&no = 260077&r (accessed November 22, 2005).

5. "Supporters of Hwang Woo-suk," audio recording, Donga.com, December 7, 2005. http://donga.com/srv/service.php3?bicode = 040000&biid = 2005120721581 (accessed December 5, 2005).

6. Yonhap News World Service, "Cloning Expert Admits Using Ova from Junior Researchers," November 24, 2005. http://bbs.yonhapnews.co.kr/ynaweb/printpage/EngNews_Contents.asp (accessed November 24, 2005).

7. "PD Diary to be Discontinued," audio recording, Donga.com, December 8, 2005. http://english.domga.com/srv/service.php3?bicode = 040000&biid = 2005120838048 (accessed December 8, 2005).

8. Hyun-joo Jin, "Scientists Divided Over Whether, How to Verify Hwang Research," *The Korea Herald*, December 9, 2005. http://www.koreaherald.co.kr/SITE/data/html_dir/2005/12/09/200512090038.asp (accessed December 9, 2005).

9. Special Reporting Team, "Hwang's Colleagues Call for Internal Probe," *Joon Ang Daily*, December 9, 2005. http://joongangdaily.joins.com/200512/08/200512082211127779900090409041.html (accessed December 8, 2005).

10. Nicholas Wade, "New Criticism Rages Over South Korean Cell Research," *New York Times*, December 10, 2005. http://www.nytimes.com/2005/12/10/international/asia/ http://query.nytime.com/search/query?ppds = byIL&v1 (accessed December 10, 2005).

11. Hyun-joo Jin, "SNU to Probe Hwang's Research Next Week," *Korea Herald*, December 13, 2005. http://Koreaherald.co.kr/servlet/cms.article.view?tpl = print&sname = National&img = / (accessed December 13, 2005).

12. Reuters, "U.S. Scientist Further Questions Korean Clone Study," December 13, 2005.

13. Ibid.

14 Mark Henderson, *"I Faked my Cell Research, Admits Cloning Pioneer," Times* online, December 16, 2005. http://timesonline.co.uk/printFriendly/0,,1–23529–1934388–23529,00.html.

15. Hankooki.com, "Cell-Making Process Can Be Repeated," *Korea Times*, December 16, 2005.   http://times.hankooki.com/service/print/Print.php?po = times.hankooki.com/ 1page/tech/200 (accessed December 16, 2005).

16. Ian Sample, "Stem Cell Pioneer Accused of Faking all his Research, Apart from the Cloned Dog," *Guardian Unlimited*, January 11, 2006. http://www.guardian.co.uk/genes/ article/0,2763,1683735,00.html.0000000000000000 (accessed January 11, 2006).

17. Dong-shin Seo, "Hwang Misappropriated $7 Million in Research Funds," *Korea Times*, February 6, 2006. http://times.hankooki.com/service/print/Print.php?po = times. hankooki.com/1page/200602/kt (accessed February 6, 2006).

18. NewScientist.com, "Cloning 'Pioneer' Donated Research Funds to Politicians," February 6, 2006. http:/www.newscientist.com/article.ns?id = dn8682&print = true   (accessed February 6, 2006).

chapter eleven

1. Etain Lavelle, "Blair Lures Stem Cell Talent to U.K. as Bush Ban Stalls Science," Bloomberg.com,   February   1,   2006.   http://www.bloomberg.com/apps/news?pid = 71000001&refer = uk&sid = aRxEPb2TKVy0 (accessed January 31, 2005).

2. U.S. Department of Health and Human Services, National Institutes of Health, "Estimates of Funding for Various Diseases, Conditions, Research Areas" (table updated February 3, 2006). http://nih.gov/news/funding researchareas.htm (accessed February 9, 2006).

3. Lisa M. Krieger. "Two Top Cancer Researchers Turn Stanford Down Over Stem-Cell Delays," *San Jose Mercury News*, November 20, 2005.

4. Lavelle, "Blair Lures Stem Cell Talent to U.K. as Bush Ban Stalls Science."

5. Wesley J. Smith, "Another Cloning 'Breakthrough': The World's First Phony Stem Cells," The Weekly Standard.com, January 2, 2006. http://www.weeklystandard.com/Utilities/ printer_preview.asp?idArticle = 6522&R-C81C (accessed December 28, 2005).

6. Richard Doerflinger, "Cloning Chaos: Misrepresentations, Hype and Outright Lies in the Name of 'Science,'" *National Review Online*, December 13, 2005. http:// nationalreview.com/script/printpage.p?ref = /comment/doerflinger20051213082 (accessed February 24, 2006).

7. Bettyann Holtzmann Kevles, "Barely a Drop of Fraud: Why It Shouldn't Taint Our View of Science, Washington Post.com, January 8, 2006. http://www.washingtonpost.com/wp-dyn/content/article/2006/01/06/AR2006010602279_p (accessed February 24, 2006).

8. Arthur Caplan, PhD, "The End of Science? Hardly. Despite Hwang's Great Charade, Science Will March Ahead, MSNBC.com, January 3, 2006. http://msnbc.msn.com/id/ 10683107/print/1/displaymode/1098/ (accessed January 3, 2006).

9. Ibid.
10. Rebecca Vesely, "Publisher of Korean Stem Cell Research Says Fraud Is Rare," *Inside Bay Area*, January 21, 2006.
11. Ihlwan Moon, "The Cloning Crisis Clouding Korea," MSNBC.com, December 18, 2005. http://msnbc.msn.com/id/10531949/print/1/displaymode/1098/ (accessed December 19, 2005).
12. Sang-hun Choe, "Lesson in South Korea: Stem Cells Aren't Cars or Chips," *International Herald Tribune*, January 11, 2006.

index